Springer Tracts in Advanced Robotics

Volume 77

Editors: Bruno Siciliano · Oussama Khatib

Satoshi Murata and Haruhisa Kurokawa

Self-Organizing Robots

Springer

Professor Bruno Siciliano, Dipartimento di Informatica e Sistemistica, Università di Napoli Federico II, Via Claudio 21, 80125 Napoli, Italy, E-mail: siciliano@unina.it

Professor Oussama Khatib, Artificial Intelligence Laboratory, Department of Computer Science, Stanford University, Stanford, CA 94305-9010, USA, E-mail: khatib@cs.stanford.edu

Authors

Prof. Satoshi Murata
Tohoku University
Department of Bioengineering and Robotics
Graduate School of Engineering
6-6-1 Aoba-yama, Sendai 980-8579
Japan
E-mail: murata@molbot.mech.tohoku.ac.jp

Dr. Haruhisa Kurokawa
National Institute of Advanced Industrial Science and Technology (AIST)
Intelligent Systems Institute
Field Robotics Research Group
1-1-1 Umezono, Tsukuba, Ibaraki 305-8568
Japan
E-mail: kurokawa-h@aist.go.jp

Additional material to this book can be downloaded from http://extra.springer.com

ISBN 978-4-431-54054-0 e-ISBN 978-4-431-54055-7

DOI 10.1007/978-4-431-54055-7

Springer Tracts in Advanced Robotics ISSN 1610-7438

Library of Congress Control Number: 2011942597

Original Japanese edition
Jikososhiki Kikai System no Sekkeiron
By Satoshi Murata and Haruhisa Kurokawa

Typeset by Scientific Publishing Services Pvt. Ltd., Chennai, India.

Printed on acid-free paper

5 4 3 2 1 0

springer.com

STAR (Springer Tracts in Advanced Robotics) has been promoted under the auspices of EURON (European Robotics Research Network)

Foreword

Robotics is undergoing a major transformation in scope and dimension. From a largely dominant industrial focus, robotics is rapidly expanding into human environments and vigorously engaged in its new challenges. Interacting with, assisting, serving, and exploring with humans, the emerging robots will increasingly touch people and their lives.

Beyond its impact on physical robots, the body of knowledge robotics has produced is revealing a much wider range of applications reaching across diverse research areas and scientific disciplines, such as: biomechanics, haptics, neurosciences, virtual simulation, animation, surgery, and sensor networks among others. In return, the challenges of the new emerging areas are proving an abundant source of stimulation and insights for the field of robotics. It is indeed at the intersection of disciplines that the most striking advances happen.

The *Springer Tracts in Advanced Robotics (STAR)* is devoted to bringing to the research community the latest advances in the robotics field on the basis of their significance and quality. Through a wide and timely dissemination of critical research developments in robotics, our objective with this series is to promote more exchanges and collaborations among the researchers in the community and contribute to further advancements in this rapidly growing field.

The monograph by Satoshi Murata and Haruhisa Kurokawa is an English translation from a recently appeared book in Japanese on self-organizing mechanical systems. This is a relatively new area of research focusing on the realization of machines and robots, made to have a certain structure and functions, which are indeed capable to adapt to unexpected situations, such as a misuse or a break-down, and ultimately reorganize themselves. As such, the book has a wide interest for scholars in the area of autonomous distributed systems, modular design and biologically-inspired robotics.

Rich in examples and case studies, deep in the discussion of various issues in the implementation and instrumentation of self-organizing mechanical systems, this volume ambitiously aims at inspiring the designers of the next generation of robots. A very fine addition to the STAR series!

Naples, Italy
August 2011

Bruno Siciliano
STAR Editor

Preface*

The notion of *self-organization* has recently been used in a variety of areas. It refers to the phenomenon in which an entity produces its own organization or structure by itself, just as biological organisms do. For example, seeds sown in the ground will sprout and leaves will grow. Then flowers will bloom, fruit will form, and finally seeds are reproduced. By simply putting the seed into the ground, it spontaneously *becomes* these things without any outside guidance. What an intriguing and fascinating process! What, then, about machines? There has never been a machine that developed by itself from a seed. Machines are built from many parts in factories using external forces. Instructions are given from the outside for each step of assembly. In short, machines are *made* to have a certain structure.

These days, the machines we need in our everyday life have become more and more complicated. It now has become difficult to grasp every detail of these machines, and every detail of the mechanisms at work. Moreover, when those machines break down or are used in unexpected manners, the behavior of the machines can become unpredictable or in fact completely useless. In situations like this, one might hope that the machine would somehow adapt to the situation by reorganizing itself. This is where the notion of *self-organizing* mechanical systems comes in. It means changing from machines that are *made* to have a certain structure and functions, to those that *become* reorganized in a desired way.

Engineers have been learning from biology for a very long time. The drawing of flying machines by Leonardo da Vinci shows that research fields such as biomechanics and biomimetics have their origins back at least to that time. However, in order to create a machine that *becomes* properly organized, it is not sufficient to copy the appearance or the superficial mechanisms of biological organisms. It is necessary to understand the architecture and mechanisms behind the organisms that enable them to function. The theme of this book is to examine the feasibility of creating such artificial systems, namely *self-organizing robots*, the title of this book, within the limitations of current mechanical engineering. We have to consider how to construct such robots, and have to find possible applications for them. Robotics in general has many different aspects such as dynamics, fabrication, control, electronic implementation, and software. In this book, in addition to

---

* This book is an English translation of "Jiko-soshiki kikai sisutemu no sekkeiron (Designing self-organizing mechanical systems)" published in 2009 by Ohm-sha, Japan. Translation and publication were supported by Grant-in-Aid for Publication of Scientific Research Result (No. 226006) from the Ministry of Education, Culture, Sports, Science and Technology (MEXT) of Japan.

these items, a robot is viewed as an autonomous distributed system. Taking this viewpoint, what needs to be considered to create robots that possess the capability of self-organization is discussed in detail.

Here is an outline of the chapters: Chapter 1 describes the philosophy of design of self-organizing mechanical systems. The discussion here may seem a little abstract but it contains a simple introduction to the notion of autonomous distributed systems, which forms the basis of the concept of self-organization and self-organizing mechanical systems. Chapter 2 considers some examples of self-organization at various levels of hierarchy in biological systems, which is our source of inspiration when we design self-organizing machines and robots. Chapter 3 presents a history of the research related to self-organizing mechanical systems. Instead of providing a complete history, we present several selected topics suitable for grasping the flow of research in this area. Chapter 4 explains some mathematics and methodology necessary for understanding theoretical background to use self-organization phenomena. Chapter 5 demonstrates the feasibility of building a self-organizing machines by introducing an actual machine that is able to assemble itself and repair itself, which is one of the achievements of our own research. Chapter 6 discusses some examples of self-organizing robots which are known as modular robots, and gives several case studies. The reader will see that a variety of robots are being developed based on many different approaches. Chapter 7 and 8 primarily explain M-TRAN, one of the most advanced self-organizing robots at this time, which can create its own shape and robotic motion. Chapter 9 addresses various issues in the implementation and instrumentation of self-organizing robots, suggesting ideas to solve them. Chapter 10 discusses the future of self-organizing robots, especially molecular-level self-organizing robots.*

It is not required to have advanced knowledge of mathematics to read this book. The knowledge that science-oriented undergraduates has will suffice. The assumed readers of this book are students in the existing robotics discipline, including areas such as mechanics, control, electronics, and computer science, and also researchers who wish to look into the area of robotics, or who are working in other areas and have interest in applications of self-organization phenomena.

This book can be read in different ways: who wish to learn the basic concept should start from Chapter 1. Those interested in the fundamentals of self-organizing mechanical systems and robots but not the philosophical arguments may start from Chapter 5 and then continue onto Chapter 6, and so be able to grasp the overall trends in this area and the current research activities. Those interested in developing actual robots are recommended to read Chapters 7 through 9, while referring to Chapter 4 as needed if the reader encounters unfamiliar mathematics.

Here, we would like to explain the background of this book. The authors were researchers in an institute formerly called the Mechanical Engineering Laboratory (MEL) of the Agency of Industrial Science and Technology of the Ministry of

* Extra materials that compliment this book, such as movies of experiments, are available at a web page. These materials can be accessed through http://extras.springer.com/ and by searching with this book's ISBN (please make sure you enter the full ISBN number, including hyphens).

International Trade and Industry, Japan. (Kurokawa is currently affiliated with the National Institute of Advanced Industrial Science and Technology, AIST, into which MEL was incorporated, whereas Murata's current affiliation is Tohoku University.) In MEL, there was a research group on fault-tolerance at the beginning of the 1980s. The group members included Kurokawa, Toshio Fukuda (now at Nagoya University), and Shigeru Kokaji, who was substantially the leader of all the research discussed in this book. This group was one of the birthplaces of self-organizing robots in Japan, as evidenced by Fukuda's pioneering self-organizing robot CEBOT in later years. When Murata joined MEL in 1987, Kokaji had just completed his innovative distributed machine called Fractal Machine. At that time, Kokaji was engaged in constructing parallel computers and had assembled entirely by himself a parallel computer system consisting of 64 microprocessors. This computer gave Kokaji the idea for the Fractal Machine. This machine was a system thoroughly based on the principle of distributed systems and provided in many ways the model for the research that followed.

In the following years, the Synergetics Research Group was founded with Kokaji as the leader. The group was eventually recognized as one of the official research groups of MEL, and new members Kohji Tomita, Eiichi Yoshida, and Akiya Kamimura joined this group one after another. The authors would like to emphasize that many of the results presented in this book are the achievements of all the members of our group.

The authors received much help from people outside the group, including the late Kazuo Tanie at Metropolitan University of Tokyo, and Toshio Fukuda at Nagoya University, who gave us encouragement and valuable advice from time to time. We thank Kohji Ito at Ritsumeikan University that the authors could meet various researchers including the late Hideo Yuasa at Tokyo University, Kazuo Hosokawa at RIKEN, and Akio Ishiguro at Tohoku University.

We also must thank colleagues overseas, including Gregory Chirikjian at Johns Hopkins University, and Daniela Rus at MIT. The authors also had various extremely constructive research interactions, such as exchange of graduate students, with Henrik Lund at the Technical University of Denmark, Rolf Pfeifer at the University of Zurich, Wei-Min Shen at the University of Southern California, and Mark Yim at the University of Pennsylvania. Many of the photographs and diagrams that appear in this book are shown thanks to the generosity of these people.

The authors are very grateful to Motoko Takenishi of Ohmsha, Ltd. who helped us to write the Japanese version of this book, and Kazuhiko Ogawa and his colleagues at NBT Corporation for their work of translation. If it had not been for their help, this book would not have been completed.

The research described in this book has been supported almost continuously by various kinds of funding, including the special funding of the Mechanical Engineering Laboratory, the funding of the former Science and Technology Agency for basic research, the NEDO Project for Robot Development, and Grants-in-Aid for Scientific Research of MEXT. The authors would like to take this opportunity to express their gratitude to all those responsible for the funding.

Lastly, the authors strongly hope that this book will help students and researchers to understand ways of thinking about self-organizing robots, and that some of the readers will be inspired to create the next generation of these mechanical systems.

July 2011

Satoshi Murata
Haruhisa Kurokawa

Contents

1 Designing by Self-Organization 1
1.1 Reductionist Design and Its Limits 1
1.1.1 Components of Mechanical Systems 2
1.1.2 Reductionist Design Theory of Mechanical Systems 3
1.1.3 Modeling and Optimization 5
1.1.4 Problems with Reductionist Design 6
1.2 Distributed Autonomous Systems and Self-Organization 6
1.2.1 From Reductionism to Self-Organization 7
1.2.2 Distributed Autonomous Systems and Theory of Design by Self-Organization 8
1.2.3 Advantages of Self-Organizing Mechanical Systems 11
1.2.3.1 Flexibility 11
1.2.3.2 Scalability 12
1.2.3.3 Fault-Tolerance 12
1.2.3.4 Problems Caused by the Involvement of Humans 13
1.3 Types of Self-Organizing Mechanical Systems 14
1.3.1 Systems and Their Components 14
1.3.2 The Complexity, the Number of Components, and the Complexity of Connections 15
References 17

2 Self-Organization of Biological Systems 19
2.1 Hierarchy in Biological System 19
2.2 Nucleic Acids: Formation of Double Helices by Hybridization 22
2.3 Protein Folding 23
2.4 Central Dogma 24
2.5 Biological Development: Assembly at the Level of Cells 25
2.6 Biological Self-repair 29
2.6.1 Reconstruction 30
2.6.2 Physiological Regeneration 30
2.6.3 Compensatory Regeneration 31
2.6.4 Wound Healing 31
2.6.5 True Regeneration 31
2.7 Self-Organization of a Group of Individuals 32
2.7.1 Cellular Slime Molds 33

2.7.2 Social Insects 33
2.7.3 Herds of Animals 34
References 35

3 History of Self-Organizing Machines 37
3.1 Work by von Neumann 37
3.1.1 Von Neumann's Two Questions 38
3.1.2 Von Neumann's Self-reproducing Automata 39
3.1.3 Universal Automata: The Kinetic Model 41
3.1.4 Universal Automata: The Cellular Model 42
3.2 Work by Penrose 45
3.3 Mathematical Models of Self-reproduction 48
3.3.1 Langton's Self-reproducing Loop 49
3.3.2 Graph Automata 50
3.4 Physical Models of Self-reproduction 52
3.4.1 Magnet System by Hosokawa 52
3.4.2 Mechatronic Self-assembling System by Klavins 55
3.4.3 Self-reproducing System by Griffith 56
References 57

4 Basics in Mathematics and Distributed Algorithms 59
4.1 Distributed System and Components 59
4.2 Diffusion 61
4.2.1 Diffusion Equations 61
4.2.2 Gradient Field 63
4.2.3 Pattern Formation by Reaction-Diffusion System 63
4.3 Cellular Automata 69
4.3.1 Field of Diffusion 69
4.3.2 Flow Field 71
4.3.3 Game of Life 72
4.4 Distributed Algorithms 72
4.4.1 Leader Election 73
4.4.2 Spanning Tree Construction Problem 73
4.4.3 Exclusion Control 74
4.4.4 Deadlock 74
4.4.5 Reliability 74
References 75

5 Artificial Self-assembly and Self-repair 77
5.1 Methods for Self-assembly and Self-repair: Homogenous System Approach 77
5.2 Hardware for Two Dimensional Units 80
5.3 Preconditions for Self-assembly Algorithms 83
5.3.1 Unit Identifier 83
5.3.2 Method and Range of Communication 83
5.3.3 Spatio-temporal Symmetry Breaking 84

5.4 Algorithm (I) for Self-assembly 85
5.4.1 Description of the Target Configuration 85
5.4.1.1 Connection Type 86
5.4.1.2 Distance between Connection Types 87
5.4.1.3 Description of the Target Configuration Using Connection Types 87
5.4.2 Strategy for Self-assembly 88
5.4.2.1 Difference Measure 88
5.4.2.2 Movable Type 89
5.4.2.3 Diffusion Field 90
5.4.2.4 Activation Criteria 90
5.4.3 Simulations and Experiments 91
5.5 Algorithm (II) for Staged Self-assembly and Self-repair 92
5.5.1 Logical Type and Description Matrix 94
5.5.2 Onion Method 94
5.5.3 Simulation of Self-assembly (Algorithm II) 97
5.5.4 Simulation of Self-repair (Algorithm (II)) 98
5.5.4.1 Detection of the Loss 98
5.5.4.2 Retrogression Signal 98
5.5.4.3 Retrogression of the Stage 98
5.6 Cellular Automata Model 102
References 103

6 Prototypes of Self-Organizing Robots 105
6.1 Classes of Modular Robots 105
6.2 Lattice-Type and Chain-Type 106
6.3 Constraints in Hardware Design for Lattice-Type Modules 107
6.3.1 Limited Space for Design 107
6.3.2 Symmetry 108
6.3.3 Degrees of Freedom for Mobility 108
6.3.4 Connectors (Connection Mechanisms) 109
6.3.5 Actuators 110
6.4 Prototypes of Modular Robots 110
6.4.1 CEBOT 110
6.4.2 Truss-Type: Fractal Machine 112
6.4.3 Truss-Type: TETROBOT 114
6.4.4 Lattice-Type: Metamorphic Robot 115
6.4.5 Lattice-Type : Crystalline 116
6.4.6 Lattice-Type: Micro Modules 116
6.4.7 Lattice-Type: CHOBIE 116
6.4.8 Lattice-Type: Three Dimensional Universal Connection System 117
6.4.9 Lattice-Type: Molecule 120
6.4.10 Lattice-Type: ATRON 121
6.4.11 Lattice-Type: Molecube 123
6.4.12 Chain-Type: PolyPod and PolyBot 124

6.4.13 Chain-Type: CONRO and Superbot 125
6.4.14 Lattice-Type: Catom 126
6.4.15 Amorphous-Type: SlimeBot 127
6.5 Hybrid Type Combining Lattice and Chain 128
References 129

7 Robotic Metamorphosis 131
7.1 System Design 131
7.1.1 M-TRAN Module 131
7.1.1.1 Shape and Function 131
7.1.1.2 Characteristics of the Shape 134
7.1.2 Basic Motions 134
7.1.2.1 Motions on the Ideal Plane 134
7.1.2.2 Constraints 136
7.1.3 Polarity 137
7.1.4 Universal Assembly and Self-reconfiguration 138
7.2 Planning Metamorphosis Procedure 139
7.2.1 Search for Metamorphosis Procedures 140
7.2.1.1 Reconfigurability 140
7.2.1.2 Exhaustive Search 140
7.2.1.3 Heuristics 141
7.2.2 Metamorphosis between Mobile Robot Configurations 141
7.2.2.1 Parallel Quadruped Form 141
7.2.2.2 Other Metamorphoses 144
7.3 Distributed Metamorphosis 144
7.3.1 Distributed System and Grouping 144
7.3.2 Meta-modules Simulating Virtual Modules 146
7.3.3 Regular Structures 148
7.3.3.1 Type-I Linear Regular Form 150
7.3.3.2 Type-II Linear Regular Form 151
7.3.3.3 Planar Regular Form 151
7.3.3.4 Three Dimensional Regular Forms 151
7.3.3.5 Cluster Flow 151
7.3.4 Motions of Planar Regular Structures 155
7.3.4.1 Basic Motions 155
7.3.4.2 Tile Model 156
7.3.4.3 Cellular Automaton 159
7.3.5 Distributed Metamorphosis by the Cellular Automaton Model 160
7.3.5.1 Cluster Flow on Planar Structure 160
7.3.5.2 Collision Avoidance, Deadlock Avoidance, and Global Consensus 162
7.3.5.3 Porous Structure 165
7.4 Various Metamorphoses 166
7.4.1 Generation of Robots from Regular Structures 166

7.4.2 Docking and Merging 166
7.4.3 Self-replication....168
7.5 M-TRAN Colony....169
References....170

8 Self-Organization of Motion 173
8.1 Robot Motion Control 173
8.1.1 Manipulator End Point Control.... 173
8.1.2 Legged Walking Robots.... 175
8.1.3 Whole Body Locomotion.... 179
8.1.3.1 Rolling Motion 179
8.1.3.2 Crawler 179
8.1.3.3 Traveling Wave on a Serial Link.... 180
8.1.3.4 Motions of Snakes and Fish.... 180
8.1.3.5 Combinations of Traveling Waves 182
8.1.4 Design of Motion Control Systems.... 183
8.1.5 Distributed Motion Control of Modular Robots.... 184
8.2 Coupled Oscillators 184
8.2.1 Synchronization by Diffusion 184
8.2.2 Entrainment.... 187
8.2.3 How to Introduce Phase Offsets.... 191
8.3 Motion Control Using Coupled Oscillators.... 193
8.3.1 Connection with Physical Systems 193
8.3.2 Global Entrainment.... 194
8.3.3 Neural Oscillator.... 195
8.4 Genetic Algorithm 197
8.5 Motion Control of the M-TRAN Robots 199
8.5.1 CPG Control System.... 199
8.5.2 Fitness and Dynamics Simulation.... 200
8.5.3 GA Optimization.... 201
8.5.4 Optimization Results and Playback Experiment.... 203
8.5.5 Real Time CPG Control.... 203
8.5.6 Issues of CPG Control.... 207
8.6 Remark 208
References.... 208

9 Hardware and Software 211
9.1 Hardware 211
9.1.1 Structure and Mechanism....211
9.1.2 Connection Mechanism....214
9.1.2.1 Magnetic Connection Mechanism....214
9.1.2.2 Mechanical Connection Mechanism 217
9.1.3 Circuitry 219
9.1.3.1 Multiple CPU System....219
9.1.3.2 Communication between Modules 221
9.1.3.3 Power....223

9.1.4 Optional Modules ... 223
9.2 Software ... 223
9.2.1 M-TRAN Simulator ... 223
9.2.2 Onboard Program ... 225
9.2.3 Program for Centralized Metamorphoses ... 225
9.2.3.1 Transformation Procedure Data and Master-Slave Control ... 225
9.2.3.2 Configuration Recognition and Role Assignment ... 226
9.2.3.3 Symmetric Conversion ... 228
9.2.4 Program for Distributed Metamorphosis ... 230
9.3 Errors and Reliability ... 230
9.3.1 Dimension Error ... 231
9.3.2 Structural Deformation ... 231
9.3.3 Dealing with Errors ... 232
References ... 233

10 The Future of Self-Organizing Robots ... 235
10.1 Challenges for Self-Organizing Robots ... 235
10.1.1 Module Size ... 235
10.1.2 Number of Modules ... 236
10.1.3 Choice between Self-reconfiguration and Self-assembly ... 236
10.2 From Mechatronics to Molecular Machines ... 237
10.2.1 Molecular Machines Based on DNA Nanotechnology ... 238
10.2.2 Self-assembly in DNA Nanostructures ... 239
10.2.3 DNA Logic Gates ... 241
10.2.4 DNA Sensors and DNA Actuators ... 241
10.3 From Nanotechnology to Molecular Robotics ... 242
10.4 Emergence of Hierarchy: The Ultimate Problem ... 245
References ... 246

Subject Index ... 247

Chapter 1
Designing by Self-Organization

Abstract. When designing mechanical systems, the method people normally use is based on what is called reductionism. Reductionism is very simple and powerful, because it is based on the one-to-one correspondence between the required functions and the necessary components of the system. Further, reductionism and the concept of centralized control are naturally suited to each other, because relations between functions and components are clearly seen in centralized control systems. The theme of this book is the theory of design utilizing self-organization, which is the opposite of reductionist design. In designing by self-organization, rather than assigning functions directly to components, *relations between components* are specified so that the components will organize the whole structure by themselves and consequently let the functions of the whole system emerge. A system built in this way does not have a central component, but instead has decentralized, distributed structure. Although it might seem a very roundabout way of building an entire system, it is a key to progress beyond the limits of reductionist design, by which it is increasingly difficult to meet demands for increasing complexity and scale. Engineering is a discipline which studies methods for creating useful things. The theory of design by self-organization is nothing other than an attempt to establish a new conceptual basis for creating such methods.

1.1 Reductionist Design and Its Limits

A systematic methodology is fundamental for the construction of not only mechanical systems but also large and complex artifacts in general [1]. The most common such methodology is based on a general scientific method, so-called Cartesian *reductionism*, which claims that you can understand a complex object by decomposing it into components so simple that they cannot be further decomposed and are very well-understood, then reassembling the components into the whole. Of course, designing an artifact emulating the function of some object (which we will refer to as a *system*) includes determining *how to build* it, and therefore, simply decomposing the object is not sufficient. If the function required of the system is simple, there often exists an obvious solution comprising a mere combination of several well-understood components. When the required function is not simple, however, the number and types of necessary components are not obvious. In the latter case, you have to decompose, or *reduce*, the required function into simpler

S. Murata and H. Kurokawa: Self-Organizing Robots, STAR 77, pp. 1 – 18.
springerlink.com

functions which are immediately realizable by simple components. Once the necessary components are determined, you can assemble them into the system originally intended. If the resulting system does not satisfy the requirements, something is wrong with either the way the decomposition was done, or the way the components were assembled. In either case, you need to make some adjustment and start again. In short, designing a system is a process of finding how it should be built while going back and forth between the opposite tasks of decomposition and reassembly. We call this "reductionist design" methodology.

1.1.1 Components of Mechanical Systems

In this book, we define a *mechanical system* as "a collection of physical components that are driven by energy from external or built-in sources, among those components there being either mechanical interaction, communication, or both, resulting in relative motions among components and changes in the internal states of components."

Since components of a mechanical system are concrete physical entities, a system can only contain a finite number of components each with a specific size, and therefore the whole system has finite dimensions. Consequently, there is a distinction between the inside and outside of a system. The outside of a system is called an *environment*. It refers to the surroundings of the system, and sometimes includes humans and other mechanical systems.

Components of mechanical systems include not only typical machine parts such as gears and screws, but also electric parts such as transistors, wires and motors, and also microprocessors for information processing. Each of these components must have clearly defined explicit or implicit input-output correspondences, in which specific input (action upon a component by other components or the environment) results in specific output (action upon other components or the environment).

Components of mechanical systems fall into the following five categories:

1. Structural components: components like a chassis or container that define the structure and the shape of the system.
2. Sensors: components that detect the state of the outside of the system (the environment), convert what is detected into information, and convey it to other components inside the system. They may receive signals from information-producing components in other systems.
3. Actuators: components that receive information from other components and act directly on the external environment according to this information. Actuators include components that generate force and motion, emitters of sound and light, and generators of heat and chemical substances.
4. Information processing components: components that receive signals from sensors and other information processing components and that perform certain computations. Some have internal states (memories) for computation purposes.
5. Power sources: energy sources for the actuators.

A mechanical system is made of components of these five types connected in a certain way. The problem of designing a mechanical system is that of selecting appropriate components and specifying their connection.

1.1.2 Reductionist Design Theory of Mechanical Systems

Designing is a process of creating something new. Therefore, the starting point of designing always has to be the question why you need to create something new. If one doesn't know the purpose clearly, one cannot develop a good design. On the other hand, if the purpose is clear and valuable, many people will get involved, resulting in outstanding work. We next give a simple review of what is involved in the process of designing.

The thought you had that initially made you want to create will give you a vague perspective on how the object or system should be realized. Based on this perspective, the process of design begins. First, the specifications of the system as a whole with its desired functions are decided; this will determine the target you want to achieve.

Next, you have to put together a construction plan. The result of this construction is then simulated and results are checked. If they are satisfactory, then the actual production begins. Finally, the actual resulting products are tested to check if they meet the target criteria (Fig. 1.1).

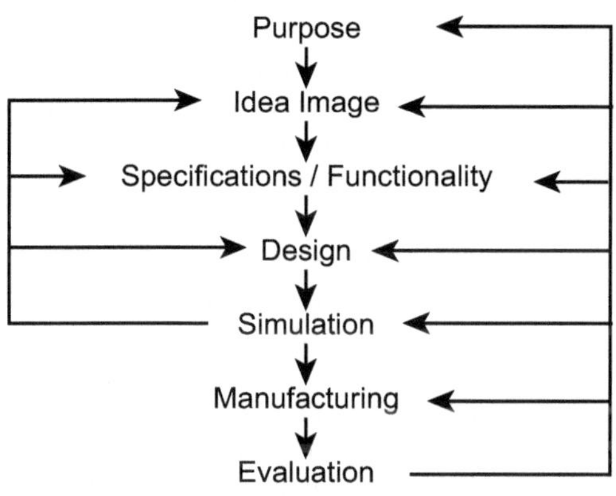

Fig. 1.1 Process of artifact creation

In reality, these products rarely meet the required specifications perfectly at the first trial, which means that the whole or a part of this process has to be repeated. By going through this cycle several times, the design becomes more and more refined. Sometimes it may be necessary even to step back to the starting point, as far back as modifying the original system specification.

Next, we focus on the 'designing' phase before the production. This phase can be divided into the following three steps:

1. Conceptual design, determining the overall specifications of the system.
2. Baseline design, deciding the basic structure and the shape of the system needed to realize the required specifications.
3. Detailed design, deciding the sizes, shapes, and materials of all the parts.

In this way, we decide the properties and the structure of our system, proceeding from abstract to concrete, from whole to part (Fig.1.2).

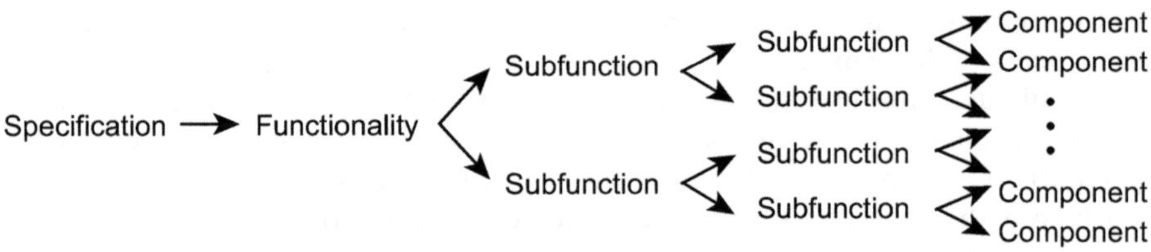

Fig. 1.2 Reductionist design. Arrows indicate decomposition; the reverse direction indicates synthesis.

Of course, it is an oversimplification to think that designing is the reduction (break up) of the required specifications into the functions of constituent parts. The first step is reduction into parts after you find out as thoroughly as possible what components are necessary, and the next is to assemble those components to build a system. There are many different ways to assemble the components that realize the same function, and there are differences in the functions exhibited by components that are assembled differently. Therefore, even after components are assembled, one needs to evaluate the system to see if it exhibits the intended functions. This evaluation involves analysis of a complex system, to which the approach of ordinary science, i.e., reductionism, is applicable. If the desired functions are not exhibited, there is some problem with the way the components were assembled. Modifications thus must be made, and then evaluation and analysis must be done again. We call this methodology *reductionist design*. (Although what is generally called designing is almost always based on this methodology, in this book we explicitly use this term to contrast it with the notion of designing by self-organization introduced later in the book.)

An important rule for reductionist design is to *achieve one-to-one correspondence of functions and components*. In other words, it is decided that a particular component realizes a particular function, and no other function should depend on that component. If a component is used for various functions, any change in the design to change a function related to that component will have undesired effects in the system. The key to successful reductionist design is to use a component for only a single function. If this is satisfied, then deciding the assembly of components will automatically determine the functions of the whole system.

1.1.3 Modeling and Optimization

There is great advantage in dividing the design problem into sub-problems. Most past and current artificial complex systems have been realized through this reductionist design. Dividing a problem, however, can cause other difficulties such as maintaining balance among different functions.

There is no limit to the possible system configurations that satisfy the requirement specifications. One has to choose one from them what seems to be the best with respect to certain criteria (*evaluation criteria*). Simply stated, determining the system configuration has two phases. The first is constructing the system framework, also called the *design model*, which determines the way components are connected, or the relative positioning of the components (*topology*). The next is deciding the values of the parameters for this model, also called the *design variables* of the model.

Therefore, solving a design problem involves deciding on a design model and then determining the optimal design variables for the given evaluation criteria. In the case of designing a vehicle that travels on the ground, for example, one first has to decide which design model to use; whether the vehicle will be a machine with four legs like a horse, or six legs like an insect, or with wheels. (Note that the design model cannot be determined by reductionist methods; the design model cannot be found by reducing the targeted functions to subfunctions).

The process of design variable optimization is as follows: first, construct a model (descriptive model) for a quantitative evaluation of how well the system satisfies the evaluation criteria. Various types of descriptive models can be used; for example, a mathematical formula like an equation of motion, a flowchart describing the state transitions and input-output relations, or in some cases even an actual physical model, so long as they can exhibit the behavioral characteristics of the target. The problem of predicting the behavior of a system quantitatively is called a *forward problem*.

For simple mechanical systems, finding an analytical descriptive model is sometimes sufficient for understanding the behavior of the system completely. In such cases, the behavior of the system can be evaluated when the system is expressed as a model formula incorporating design variables, and the variables which optimize the evaluation functions are uniquely determined. This completes the design. In most cases, however, an optimal solution cannot be obtained analytically, due to the issues such as the complexity and non-linearity of the system, the complexity of the environment, or the complexity of the evaluation criteria. In those cases, either analytical evaluation of an approximate model that partially represents the system behavior, or *simulation* that reproduces the system behavior on a computer will be employed.

There are several kinds of simulations, such as discrete event simulation, dynamic simulation and probabilistic simulation. When the system is simple enough, the entire system can be represented by a single simulation model. However, in general, the system is first divided into several smaller parts, i.e., subsystems, each of which is simulated independently. This makes the input-output relation of each subsystem clear. A model of the entire system is then constructed with those subsystems used as building units.

1.1.4 Problems with Reductionist Design

Even in the case of a very large and complicated system, by employing the reductionist approach, the task of design can be reduced to tasks of designing smaller systems that can each be handled by a single designer. System parts defined in subtasks are connected and then a simulation of the entire system is done to predict its behavior. Optimization of the system sometimes entails repetition of the cycle of decomposition of a system.

In this manner, the reductionist theory is very powerful in designing artificial systems. One should not forget that it has certain limitations, however. For example, it is known that when a system has some non-linear property, a small difference in the initial values sometimes causes a huge deviation in the course of time (butterfly effect). This means even when a system is expressed accurately by a descriptive model, its behavior cannot be simulated completely. Recent advances in computing technology have enabled simulations of complex non-linear descriptive models, but it is fundamentally impossible to predict completely the behavior of non-linear systems. What is required instead is a methodology for verifying or guaranteeing the most plausible behavior of systems.

There is another limitation in the reductionist approach. Think of the functions of the human brain, such as consciousness and intelligence. They are not determined by the intrinsic functions of the components, neurons, but rather by their *interactions* with each other. This means that, even if one divides a brain into pieces, one never obtain a component that determines human will. This example does not refute the reductionist theory, but it is evidence that investigating the components does not solve all the issues.

1.2 Distributed Autonomous Systems and Self-Organization

We mentioned earlier that deciding on the design model is an issue beyond the scope of reductionist design. We have to note that deriving a design model for complex systems itself is a very important and difficult problem. Moreover, the systems of our interest are shifting from centralized ones that are suitable for reductionist design theory to decentralized, distributed ones in which many of the same components scattered over the space (the World Wide Web is a typical example). In many conventional systems, components such as information processing units and actuators have to be concentrated in one place. But nowadays, it is often more efficient to distribute components in different locations, thanks to small, high performance components such as motors, microprocessors, and sensors, and progress in network technology. In such systems, since distributed components collectively realize the function, correspondence between the component and the function is not certain any more. Namely, reductionist approach does not work effectively in these systems.

Since the complexity and the scale of mechanical systems these days are rapidly increasing, they will soon reach a point where they are impossible to create with reductionist design methodology. We do not yet have a clear solution to that problem, and the current situation almost gives one the feeling that there is no way out. In order to deal with this situation, it is natural to seek a new conceptual approach, a methodology based on a new principle which is different from the existing design theory, i.e., the reductionist one. If engineering is *a discipline for establishing methods* for creating things, finding a new methodology for designing systems is indeed a central theme for engineering.

In this section, we introduce the theory of design by self-organization, which is the opposite of the reductionist design theory. The key ideas here are *autonomous distributed system* and *self-organization*. These terms refer to multiple autonomous components being distributed, and then *organizing themselves*, creating an integrated structure.

1.2.1 From Reductionism to Self-Organization

In the reductionist design, the functions desired for a system are reduced to the particular functions of each of its components, and then the combinations of the components are optimized. As we saw in the previous section, however, designing complex or non-linear mechanical systems exposes the limitations of such a reductionist approach.

One possible solution to this is utilizing self-organization, which is the theme of this book. It is the extreme opposite of the reductionist design; if you call the reductionist design *engineering to do something*, then design utilizing self-organization should be called *engineering to let things become something.*

In the theory of designing by self-organization, instead of reducing the required functions directly to particular components, the specifications of components are left undefined. These components are expected to *organize themselves* until they form a system that possesses the desired functions. This idea is close to selective breeding, the process for improving crops and farm animals in agriculture. The more complex the system is, the more difficult it is to logically determine the optimum design model and variables. Instead, one should adjust the growth principles embedded in the system while observing the behavior of the entire system, thus leading the system to develop into what is desired. This can be understood as a kind of *meta-design* in the sense that you do not directly design the system to be built, but rather design the generating principles of the system.

Before starting discussion of the process of design by self-organization, let us explain a little about the term *self-organization*: it refers to the property of a system to create organization and structure by itself without external help. In other words, it is a process of generating order out of randomness. It was the philosopher Kant who first used this term, and after that it became widely known and used in physics. Although the term may seem to have a clear intuitive meaning, it is difficult to give it a precise definition, as it is used with different meanings depending on the context.

In physics, it is usually used to describe two phenomena. One is the phase transition from the state of disorder to that of order. Examples include ferromagnetic transition and crystallization, both of which can be explained by the minimization of the free energy. These are phenomena of closed systems, and can be called static self-organization. The other is structure formation in non-equilibrium systems, order formation in dynamical systems where matter and energy are moving in and out. Pattern formation and rhythm generation in reaction-diffusion systems are typical examples. Prigogine's dissipative structures [2] and Haken's synergetics [3] are theories explaining these phenomena.

In chemistry, there are dynamic reactions such as reaction-diffusion systems and oscillating reactions by autocatalysis, but the term self-organization is usually used for static self-assembly, such as molecular self-assembly, colloidal crystals, and self-organizing monolayers.

In biology, the term is used rather broadly to cover various levels of hierarchies in living organisms [4, 5]. Examples range from molecular self-assembly such as folding of proteins and formation of lipid bilayer membranes, to dynamic phenomena such as homeostasis, embryogenesis, and behavior of animal groups. The term is sometimes used even in economics and social science; it has become a convenient expression to use when people want to apply the phenomena described above to humans and society.

The term self-organization is used very broadly as seen above, and we also use it in this book without clearly defining its meaning. As engineers, we should learn from the examples of self-organization found in physics, chemistry and biology, and use those insights to establish a new methodology.

1.2.2 Distributed Autonomous Systems and Theory of Design by Self-Organization

The idea of designing by self-organization originated from research into *distributed autonomous systems*[1]. To be more precise, the theory for designing distributed autonomous systems is the theory of design by self-organization. We chose the latter name in this book, because it describes the design method rather than the type of system.

A distributed autonomous system is defined as a system without any particular component which directly controls the entire system, each of its components (individually or as subsystems) realizing the functions of the system as a whole, through cooperative or competitive interaction with each other.

There are two requirements for a system to be a distributed autonomous system. One is the autonomy of individuals. Here an *individual* refers to a component that can be regarded as an independent entity[2]. For example, if we consider the automobile traffic as a system, each automobile is an individual constituting the system. In a distributed autonomous system, each individual takes actions on its own accord. That is, each individual has the capability for recognition, assessment and

---

[1] The notion of distributed autonomous systems can be found in various research areas. In this book, the term refers to the idea proposed in the course of the research project (1990-1993) in Japan supported by the Grant-in-Aid for Scientific Research, the Ministry of Education, Science and Culture [6, 7].

[2] In distributed autonomous systems, the term "individual" is often used instead of "component." The rationale is that, since an autonomous component is supposed to be an entity with appropriate capability for making its own judgment, it should be regarded as not just a component or part but an individual entity.

action, and decides its behavior based on its own evaluation criteria. The other requirement is the capability for self-organization. In distributed autonomous systems, interconnections and interactions of individuals are not fully specified a priori, and thus there is a certain freedom of action. A state of a system consisting of many individuals changes from a random state to a state with some particular order. This is the process of *self-organization*, which enables flexible adaptation to changes in the environment or changes in the specifications.

Such a system does not work if we do not know the outcome of the self-organization process. Thus, the main issue in design by self-organization is how to control the self-organization process so that it will result in the desired functionality. In the example of automobile traffic, if all the cars simply try to arrive at their destinations as quickly as possible, the roads will be heavily congested and consequently each car will take a much longer time to reach its destination. Instead, what is needed is to introduce a set of rules, such as that cars turning to the left having right-of-way, to facilitate the entire traffic flow so that the efficiency of the whole traffic system is improved. The basic principles underlying the behavior of individuals must *generate order in the entire system.* If we set these rules appropriately, it becomes possible to direct the self-organization of the individuals toward the desired behavior.

The design starts with determination of the specifications of components (individuals) and their relations, or the rules for interactions among the components. As opposed to reductionist design, in design by self-organization, the specifications of the components are not deduced from the functions to be achieved. The functions of the whole system are evaluated for the first time when the results of interaction among the components are revealed. Therefore, in designing the components and then the relations among them, it is important that they be given the potential for realizing the desired functions of the whole system.

This potential may seem to be a vague concept. One may get an idea of this by considering building with blocks; if there is only one way of connecting blocks allowed, this automatically limits the things that can be built using those blocks. Deciding upon the components and their relations is also analogous to establishing axioms in mathematics. The theorems that can be proved from the axioms represents the functions that can be exhibited by a system. Once the axioms are assumed, the set of theorems it can prove is automatically determined. Similarly, if the specifications of components and the relations between the components are not chosen well, the desired functions will never be obtained.

How then can these components and rules be determined correctly? At present, there is no way other than trial and error. One has to set provisional specifications and types of relations for components and then observe the self-organization of the system, either by simulations, or by experiments with real components. This is the only way to evaluate the tentative design. The results of evaluation are the basis for the next modified design of components, based on which another

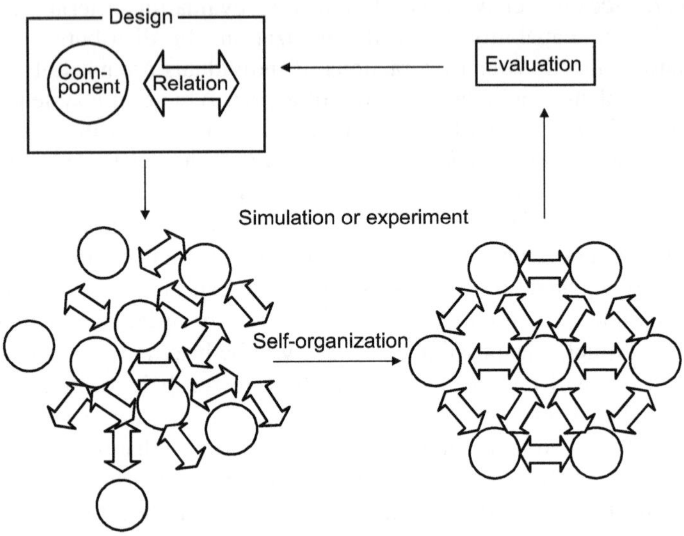

Fig. 1.3 Design by self-organization

self-organization will be carried out and tested. These processes are repeated until a satisfactory design is obtained (Fig. 1.3).

Even in reductionist design theory, evaluation by simulations or by experiments after assembling components into a system is an indispensable step. In that case, if evaluation reveals some shortcomings in the design, the components causing the trouble can be directly identified and adjusted. In design by self-organization, however, it is not immediately obvious how to make adjustments because it is not always possible to know how components are related to each other in advance, and moreover, any change in component specifications or their way of relating to each other will always affect the whole system.

To put it differently, with design by self-organization, as long as it is possible for the completed system to have the required functions, there must be a certain leeway in the original configuration. This freedom allows it to be somewhat ambiguous which component will become which part of the system, so the components must have versatility (flexibility) of function (Fig. 1.4).

Another advantage of design by self-organization is that if you know the effect of changes in the rules and resulting change of the system function in advance, it is possible to create a system that detects any change in the external or internal situation during operation and adjusts its functions flexibly to cope with the change. Capacity for this kind of dynamic self-reconfiguration is a distinctive feature of self-organizing systems.

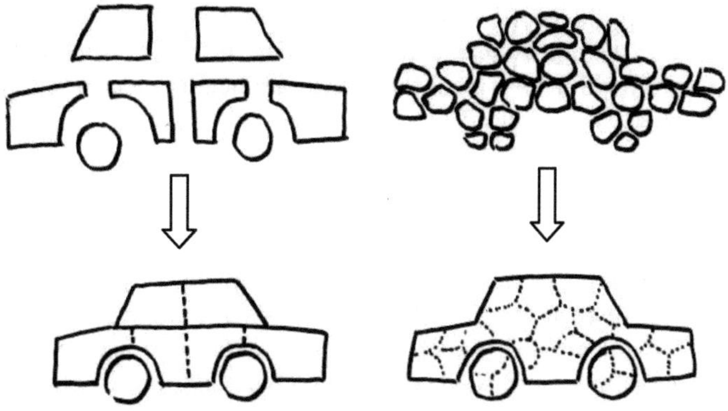

Fig. 1.4 Reductionist design (left) and design by self organization (right).

1.2.3 *Advantages of Self-Organizing Mechanical Systems*

Self-organizing mechanical systems have advantages over conventional mechanical systems. Let's look at them one by one.

1.2.3.1 Flexibility

In a self-organizing mechanical system, a function is a consequence of interaction among many components. It is possible for the same set of components to achieve different functions by changing the type of possible relations among the components. This means that one mechanical system is capable of a wide variety of functions, or in other words, it has *flexibility*.

In the reductionist design, one defines a fixed model so that it realizes the required functions satisfactorily. The functions of the system do not expand beyond what is allowed by the model. Multiple functions can be realized only by equipping the system with each of the functions in advance, just like a Swiss army knife.

On the other hand, in self-organizing mechanical systems, by carefully choosing a set of a few types of components, their interconnections can result in a wide variety of functions, such that a good set of building blocks can form a variety of interesting structures. These components are called *modules*. Modules are components of a system whose inputs and outputs are limited and standardized, so that the connections with other modules are guaranteed to be within the range defined by that standard. A system can be constructed by combining many modules.

A flexible system may have as many functions as the number of combinations of its modules. On the other hand, there are overhead costs incurred by packaging and wiring each module, and also there is a limit to the possible functions due to the standardization, therefore making it difficult to achieve the optimal configuration of the whole system. Consequently, modular systems are not suitable for systems where optimization is critical, but are a practical choice for situations where a reasonable variety of functions is required.

1.2.3.2 Scalability

The functions of self-organizing mechanical systems are dependent upon the number of its components. *Scalability* of the system is the system's adaptability to an increase or decrease in its size[3]. Since the cost for building a large scale system is substantial, it is desired that such a system can adapt to use conditions that are not optimal, that is, it should be versatile to some extent. For example, although all the characteristics that users have in common must be recognized at the start of the design stage, once the system is in operation, it may turn out that the actual users have demands slightly different from what was assumed. To deal with such change in demands, large-scale (mechanical) systems should have the capability to expand its functions or remove unnecessary functions as needed.

Typical examples of system expansion include adding memory and hard disks and upgrading CPUs of computing systems[4]. To do this, we have to shut down our PC, but a large scale system often needs to be kept running while components are added or deleted. In the Internet, which is the largest of all the systems ever created by humans, computers are added and removed perpetually.

1.2.3.3 Fault-Tolerance

Scalability gives the system the ability to continue its operation even when part of the system stops functioning, the remaining components maintaining operations by themselves. It is not easy for a large scale system to guarantee reliability of all the components, because the number of components comprising the system is enormous. In order to secure reliability of components under certain cost constraints, systematic checkups and overhauls should be regularly carried out according to the system's service life and according to the conditions of operation. This does not resolve all the issues, however.

In conventional design of mechanical systems, there are many ways to deal with system failures. First of all, in order to avoid failures as much as possible, the reliability of components should be improved. Next, the system as a whole should be set up in such a way that failures in some components do not cause immediate fatal breakdown of the system. This is the idea of *fault-tolerant design*. In fault-tolerant design, the system should have *redundancy*, so that when one of its components fails, there are other components to replace the function of the failed component. In particular, when there are multiple components working in parallel

---

[3] The term extensibility is used with the same meaning.

[4] In such systems which only deal with information, one says that the system is scalable when the throughput (the speed of information processing) of the whole system is proportional to the amount of resources (such as the memory size). In general, it is difficult to make the amount of used resources and the achieved function linearly proportional.

provide the same function (*duplication of components*), a failure in one component decreases the efficiency but does not cause complete loss of any functions.

In addition, attempts are made to achieve *failsafe design*, which keeps a system safe in case of system failures caused by incorrect user operation or malfunctions. The system is designed to run in a special safe operating mode if components fail to function correctly or if the user operates the system incorrectly. For example, it is safer if the engine of a car is designed to stop as soon as there is an engine failure.

Information systems are typical examples of systems in which fault-tolerant measures are introduced relatively easily[5]. For example, the whole system is often duplicated by installing extra hard disks. The system is equipped with two sets of hard disks and every piece of data is written on both sets. When reading data, the read operation is judged to be successful if the data read from the both sets match; otherwise, it is judged that there has been some error. If there are three sets of hard disks, it is possible to identify which of the three is (likely to be) faulty by the majority principle. This is the idea that the control systems of space ships and the RAID system are based on.

However, simple fault-tolerant systems are always faced with the problem that redundancy is costly. Increase in cost is inevitable when redundancy is introduced, but the cost is not reflected in the efficiency of the system. The more components there are, the greater the cost for maintenance, because of the increase in the number of components that might fail. Moreover, it is very difficult to test duplicated systems. Therefore, duplication redundancy of all the components does not necessarily increase fault-tolerance of the system.

The system process in which after a failure in the system, the system does not breakdown totally but stays in operation with limited functions depending on the degree of the failure is called *graceful degradation*.

1.2.3.4 Problems Caused by the Involvement of Humans

Making a system *foolproof* is one approach to reducing human error during operation. This is a kind of the failsafe design mentioned earlier in this section, in which the system is designed so that humans cannot operate it incorrectly, or so that incorrect operations do not cause accidents.

A related serious problem is that the process of constructing a system itself is not free from the risk of human error. For the case of a very large scale system, it is not always clear that the design specification itself has no inherent problems. It is very important to verify consistency of the specifications with each other[6] before starting the actual designing process, because if there is any such inconsistency, it will be impossible to satisfy all the specifications.

---

[5] The term *dependability* is also used for various systems. It is one aspect of the quality of service provided by the system, namely, how much of the intended service of the system is provided properly. This is determined by the system's reliability, fault-tolerance and operability.

[6] In the area of designing software systems, specification description languages and system description languages are developed for such purposes.

Even when there is no inconsistency of the specifications with each other, there always will be errors in programs written by humans. The greater part of the time required for building a system including software is actually spent on removing such errors (debugging). The process of debugging is complete when the system works as defined in the specification, but it is not easy to remove all the bugs. Moreover, testing is required after debugging to see if the bug was successfully removed. It is sometimes the case that people find bugs in large scale systems during actual operation, even after substantial testing has been done.

1.3 Types of Self-Organizing Mechanical Systems

Many of the problems with mechanical systems we have discussed so far can be largely resolved if more flexibility and fault-tolerance is introduced to the mechanical systems. Design, production and maintenance all incur costs if done by humans. If mechanical systems are equipped with the capability of self-organization, they will contribute to reducing design, production and maintenance costs. Distributed autonomous systems may offer a basic framework for such systems.

1.3.1 Systems and Their Components

In distributed autonomous systems, the components are individuals, autonomous entities that have their own behavioral rules and realize the functions of the whole system by interacting with each other. Below are general aspects of the relation between individuals and systems in distributed autonomous systems [8-10]:

1. The system is comprised of many individuals (components) that are distributed over a space (distributed system).
2. Individuals are similar or the same, in the sense that they can be replaced by another (homogeneity of individuals)[7].
3. Each individual behaves autonomously (autonomy, spontaneity of individual action).
4. The system and its individuals comprise an open system that takes in energy to maintain activity.
5. Each individual carries global information on the organization of the whole system. The global order of the whole system is generated and maintained by the coordination (connection, interaction) of many individuals (order formation of the system). There are fields of interaction among individuals (field of interaction), in which the information is locally exchanged among individuals (locality of communication).
6. Interactions between individuals are not determined in advance or may change at any time, depending on the surrounding conditions (non-determinism of interactions).

---

[7] Items 1 and 2 are summarized with the word "uniform", meaning the distribution of homogenous components over a space.

7. When the purpose of the system or the environment changes, the system adapts to the change (adaptability to the environment) by adjusting its own structure.

If a system satisfies the above conditions, it can be regarded as a distributed autonomous system. The actual form of the systems varies substantially depending on what is taken to be an individual.

1.3.2 The Complexity, the Number of Components, and the Complexity of Connections

How much complexity must the components have and how many of them must be used in order to construct a self-organizing mechanical system? In this section we consider the issue of complexity and the number of components, and the complexity of the connections between components.

It is difficult to define precisely the complexity of a component; for now, we only posit that complexity is closely connected to the amount of information a component carries, for example the number of possible states the component has and what functions it has in those states. In addition, we posit that the amount of this information is determined by the physical size of the component. For example, internal states of an atom include its location and speed in a 3-dimensional space, and its spin. In the case of a molecule, in addition to those, there also are its possible spatial arrangements (conformations) and its binding to other molecules.

Possible functions of a single atom or molecule are quite limited. Even if the spin state or conformation is taken to represent information, it expresses only one bit of information. On the other hand, the number of atoms and molecules we can use is enormous; for example, 12 grams of carbon consists of 6×10^{23} (Avogadro's number) of carbon atoms. If we can control the states of individual atoms independently, it is possible to use this 12 grams of carbon as a massive memory.

While it is not quite possible yet to manipulate atoms and molecules directly, it is possible to create very minute structures using photolithography. Current semiconductor technology (as of 2010) allows construction of transistors 32 nm in size, enabling a single microprocessor to have 8×10^{10} transistors. The world's first microprocessor, the INTEL 4004, which was a 4-bit processor and had 2300 transistors, would fit within a 12 μm square with the current level of integration. This means that it is possible to make a component 10 μm in size with the computation power of a simple calculator.

Micromachines, or systems that have micrometer-size moving parts, are also being made. It is also possible in principle to combine them with computation circuits. Most of the currently available micromachines are simple ones such as static actuators or rotating wheels, but their possible integration degree has reached a million. An example of such systems is the Digital Mirror Device (DMD). These kinds of micromachines still need external power sources.

For typical meter-scale machines, we can use not only various mechanical components, but also microprocessors and micromachines described above. Some state-of-the-art humanoid robots that are equipped with many of such electronic and mechanical devices can move like real humans. However, even if super high-performance humanoids like Astro Boy become available, though, they would be very costly so that an application requiring many humanoids would be unrealistic.

To summarize, if the size of a component is small, its functionality may be limited, but many of the same components can be used because each one takes up less space and less cost. When a component is larger, it may have more functions, but fewer of such components can be used due to cost and reliability constraints.

There is also the issue of homogeneity of components. Atoms and molecules are usually obtained through chemical reactions and therefore it is almost impossible to generate them individually (on the other hand, molecules generated by chemical reactions are completely homogenous and the cost of production per molecule is extremely small). Consequently, components available for nanometer-size systems are either all the same or a mixture of a few different kinds of components.

On the scale of micrometers, it is possible to control components one by one using photolithography, and microprocessors are indeed produced in such a way. Systems that have homogenous components placed in a regular repeated pattern, like semiconductor memories, are easily mass-produced and therefore can be provided cheaply. Microprocessors and memories can have high integration/density because they are circuits with no moving parts; it is very difficult to produce highly integrated mechanical systems with actively moving components.

We have discussed the size and the number of components here, but they are not the only factors deciding the level of functions of self-organizing machines. Since self-organization is a consequence of interaction between components, the way and the density with which components are connected are also important. For example, the function of a brain is determined by the connections of neurons. A human brain has 10 billion neurons, and each neuron is connected to several thousand other neurons. The flexibility provided by this enormous number of redundant connections enables a brain to memorize, learn, and judge.

Brain cells interact using neural impulses. The interactions of cells making up biological systems are based on dissipations of chemical substances and mechanical effects such as pushing and pulling. Cells contain many different kinds of molecules for those purposes, and in order to respond to them selectively, they are equipped with receptors, channel elements and signaling pathways.

Biological cells also have the ability to communicate with remotely located cells, for example using hormones and neural systems. Numerous brain cells of the same type are connected regardless of their spatial separation through nerve fibers into a complex network, whereas effects of normal chemical or mechanical reactions are greater with closer spatial proximity. An artificial creation that is similar to the brain is computer networks such as the WWW. In such systems, the computer interactions are independent of geographical locations, and a vast amount of data is exchanged according to specific protocols.

Here, let us make a bold simplification:

> The total functional capability of a mechanical system =
> the complexity of its components × the number of components.

The complexity of the interactions is reflected in the complexity and the number of components. Of course the complexity of the interactions is in proportion to a certain power exponential of the number of components, but for now we make the above simple hypothesis. Now, using this equation as the evaluation standard, what is the size of components that maximizes the functions of the whole? (Fig. 1.5)

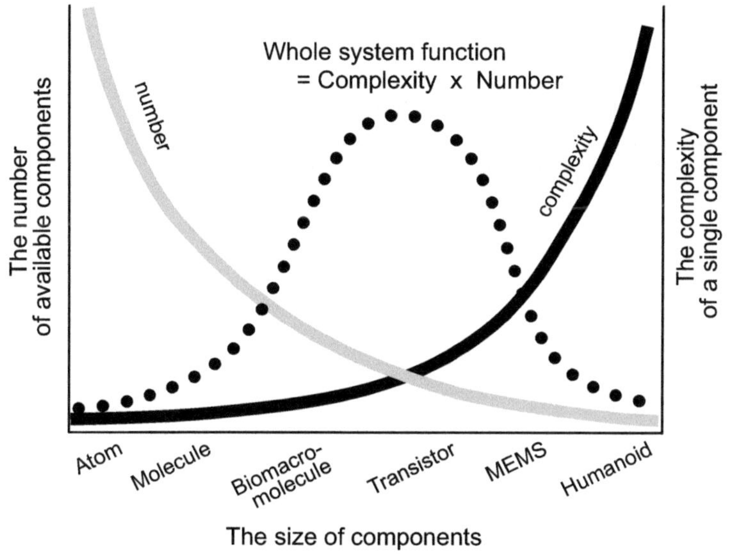

Fig. 1.5 The relation of the size and complexity of components to the functional capability of the entire system

Intuitively, an integrated circuit of about a billion transistors, each of whose size is a few tens of nanometers, seems to be this optimum component. Our current interest, though, is in building mechanical systems, so it is realistic to use components on the order of millimeters or larger, and in the following chapters, most mechanical systems are built from components (modules) on the order of centimeters. This however is simply a reflection of what components are currently available, independent of the above discussion of optimality. We will come back to discussion of the optimal size of components in the last chapter.

References

[1] Simon, H.A.: The Sciences of the Artificial, 3rd edn. MIT Press (1996)

[2] Prigogine, I., Nicolis, G.: Self-Organization in Non-Equilibrium Systems. Wiley (1977)

[3] Haken, H.: Synergetics –An Introduction. Springer Series of Synergetics, vol. 1. Springer, Berlin (1978)

[4] Camazine, S., et al.: Self-Organization in Biological Systems. Princeton Studies in Complexity. Princeton Univ. (2001)
[5] Bonabeau, E., Dorigo, M., Theraulaz, G.: Swarm Intelligence-From Natural to Artificial Systems. Oxford Univ. (1999)
[6] Ito, M., et al.: Special issue on Decentralized Autonomous Systems. J. Soc. Instrum. Control Eng. (Keisoku to Seigyo) 29(10) (1990) (in Japanese)
[7] Ito, M., et al.: Special issue on New Developments in Decentralized Autonomous Systems. J. Soc. Instrum. Control Eng. (Keisoku to Seigyo) 32(10) (1993) (in Japanese)
[8] Ito, M.: Construction of Decentralized Autonomous Systems. J Soc. Instrum. Control Eng. (Keisoku to Seigyo) 29(10) (1990) (in Japanese)
[9] Ito, M.: Future Challenges in Decentralized Autonomous Systems. J. Soc. Instrum. Control Eng. (Keisoku to Seigyo) 32(10) (1993) (in Japanese)
[10] Ito, M., Ichikawa, A., Suda, N. (eds.): Jiritsu-Bunsan Sengen (The Distributed Autonomous Manifesto), Ohmsha (1995) (in Japanese)

Chapter 2
Self-Organization of Biological Systems

Abstract. Biological creatures have abilities to generate various kinds of order in a self-organizing manner, and therefore they can be regarded as autonomous distributed systems. When we see biological organisms as autonomous distributed systems, cells are the components of these organisms, because each cell behaves autonomously as an individual entity, taking in substances and energy from the external world, carrying out its own reactions, responding to environmental changes, and even reproducing itself. Moreover, in the case of multicellular organisms, each cell differentiates from others and specializes to have particular functions in order that an assembly of cells can generate complex structures. Such an assembly can act as an autonomous individual that realizes intricate functions which cannot be carried out by individual cells. If we regard individual animals as components, herds of animals are also a kind of self-organizing system. These herds exhibit a variety of complex and clever behavior, in which distinct roles may be allotted to different members of a herd, or in which individuals of different species are in symbiotic relations. The notions of autonomous distributed systems and design by self-organization introduced in the previous chapter were born from abstraction of such biological systems. Mechanisms in biological systems are the result of four billion years of evolution, and in them we can find many clues for artificial system design. In this chapter, we introduce typical and important examples of mechanisms in biological entities.

2.1 Hierarchy in Biological System

The entire biological world can be regarded as a huge *hierarchical system*, each layer of which is a system made up of components of various sizes and complexities. A system in one layer is made up of components which in turn are systems in the layer below; *atoms* form *molecules*, molecules form *molecular machines* and *organelles*, molecular machines form *cells*, cells form *organs*, organs form *individuals*, individuals form *herds*, and herds form a *society*. Although each layer operates with a different self-organization mechanism, the hierarchy as a whole is a remarkably well coordinated system (Fig. 2.1) [1, 2].

S. Murata and H. Kurokawa: Self-Organizing Robots, STAR 77, pp. 19–35.
springerlink.com

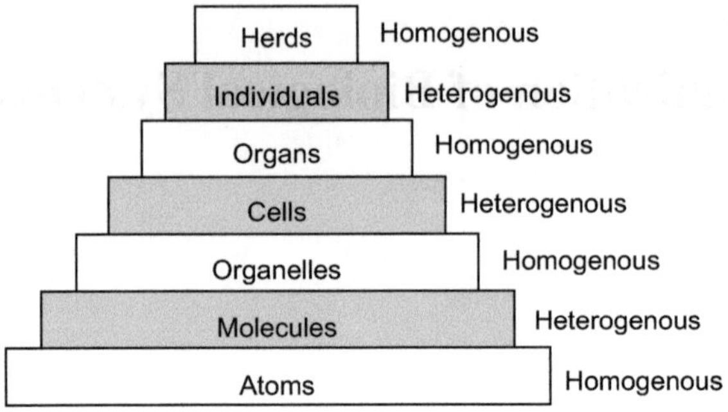

Fig. 2.1 Hierarchical structure of the biological world

Let us look at the layers starting with molecules. There are four major types of biomolecule, i.e., *nucleic acids*, *proteins*, *lipids* and *sugar chains*, which make up the bodies of biological entities. Nucleic acids are molecules for storing and transmitting genetic information. In a sense, they are the design blueprints of biological systems. Proteins provide many different functions: catalyzing of various chemical reactions in cells as enzymes, transporting substances inside and outside of cells, forming fibers and muscles to maintain body structure, etc. Lipid molecules assemble into cell membranes and sugar chains work for inter-cellular adhesion and communication.

Typical examples of self-organization at the molecular level include self-folding of proteins and hybridization of nucleic acids. Proteins are first synthesized into linear chains of amino acids, but they cannot function in this chain form; they have to be folded into particular conformations in order to function. The folding process of proteins into a certain conformation that performs a particular function such as catalysis is a result of interactions between numerous amino acids on the chain. This is an example of the self-organization processes called *cooperative phenomena*. The hybridization reaction of deoxyribonucleic acids (DNA), in which two strands of DNA joint into a double helix, is also a cooperative phenomenon.

The next layer in the hierarchy is that of organelles, such as nuclei, Golgi apparatuses, mitochondria, and chloroplasts. These are molecular machines with specific functions, produced by self-assembly of component molecules of lipids and proteins, etc. In the self-assembly process, molecules colliding with each other in solution gradually arrange themselves into a certain configuration following the principle of minimization of free energy. Typical examples of such self-assembly include the formation of a ribosome (a molecular machine that translates genetic information into proteins) from several kinds of ribonucleic acid molecules, and the formation of the cell membrane structure called lipid bilayer from phospholipids.

In the layer of cells, the situation is far more complicated. Cells must contain a certain number of organelles to survive, but the necessary number of each type of organelle contained in a cell and their activity levels must be adjusted differently for each cell and its environment. In other words, there is a need for a complex control mechanism that perpetually enhances or suppresses the functioning of the individual organelles, depending on the environment. This need is realized by the network system of gene regulation in which thousands of different genes control the expression of each other. Cellular division is one such distinctive feature of cells. Cells follow a cycle or program for division in which they replicate the whole cell, including the genetic information. This is what is called *self-replication* (or self-reproduction). Considerable numbers of the cellular level mechanisms underlying these phenomena are being elucidated by molecular biology, but it seems that it will be a long time before a systematic theory can explain the whole process.

The process of cells forming organs and organs forming individuals is called *(biological) development*. The mechanism of development is not fully understood yet, but it seems that cell division, apoptosis and differentiation required for each stage of development take place through various communications between cells; in particular, signal transmission by chemical substances. The process of *self-repair* (also called *self-healing*), a mechanism where a living individual heals itself when its body is damaged, also has a deep connection with the processes of development. As described above, there are many variations in the self-organization process of biological systems, and each layer of the hierarchy employs a different mechanism.

Now look at Fig. 2.1 again. One may note that homogeneous layers and non-homogeneous layers appear alternately in the biological hierarchy [3]. For example, the layer of organelles is homogeneous, as an organelle consists of a large number of molecules of the same kind, whereas the layer of cell is heterogeneous, because a cell contains many more kinds of organelles than the kinds of molecules that an organelle contains. Similarly, an organ consists of many homogeneous cells, while an individual is composed of different organs. You can follow the alternating homogeneous and non-homogeneous layers as you go upward in this hierarchy.

The authors think that biological entities are created through evolutionary processes, not by the Creator. However, let us assume the existence of the Creator and imagine how he went about designing biological organisms. In a non-homogeneous layer, an individual is composed of components, organs for example, each of which is in charge of a different specific function. This is clearly a reductionist design. On the other hand, homogeneous layers seem to be self-organized. Let us look at this as a lesson for engineers. Engineering has always relied on reductionist methodologies, focusing only on the non-homogeneous layers in the biological hierarchy. If we focus on those homogeneous layers, we should be able to discover a new design methodology, which is successfully used by Mother Nature.

In the following sections, we discuss some important examples of biological self-organization, with application to engineering in mind.

2.2 Nucleic Acids: Formation of Double Helices by Hybridization

Biological cells are often thought of as complicated chemical factories, because a tremendous number of kinds of protein, nucleic acid and sugar all are synthesized from simpler molecules inside the cells. In particular, nucleic acids (DNA and RNA) are made up of nucleotides — molecules consisting of bases, phosphates and sugars (See Section 2.4).

DNA molecules are important for storing genetic information. The information is encoded as sequences of four different nucleotides: adenine (abbreviated A), cytosine (C), guanine (G) and thymine (T). DNA usually exists as two long polymer chains joined together by hydrogen bonds, forming the famous double helix structure.

The nucleotide sequences in the DNA strands in a double helix are always arranged so that adjacent nucleotides of the two strands are one of the Watson-Crick complementary base pairs (A only bonding to T, and C bonding to G). In other words, if two DNA strands have Watson-Crick complementarity, they can form a double helix (Fig. 2.2). This is the process called *hybridization*.

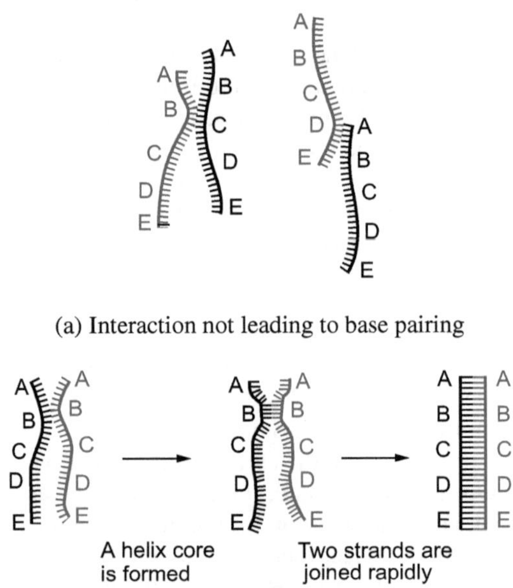

(a) Interaction not leading to base pairing

(b) Interaction resulting in base pairing

Fig. 2.2 Hybridization of DNA strands (The Cell 3rd Edition ©1994 Garland Science)

Although there are only four kinds of nucleotides, the size of actual base pair sequences is enormous; for example, there are approximately three billion base pairs in human DNA. If a small DNA segment, consisting of 20 to 30 bases, is placed in a cell, it is able to search inside the cell through diffusion to find a particular section in the DNA that has a complementary base sequence and bonds to that section. This is because it bonds only to the particular sequence that has consecutive complementary base pairs. This process is explained by a chemical reaction based on the minimization of free energy.

2.3 Protein Folding

Proteins are synthesized from twenty different amino acids (See Section 2.4). Just after the synthesis, they are linear chains of indeterminate form called *random coils*, but they rarely exist as random coils in cells; they are immediately folded into particular shapes in order to function. Self-folding of proteins is the process of transforming from a random coil into more stable structure, driven by thermal agitation; intensive collision with surrounding water molecules and parts of themselves, which replaces weak bonding of segments by bonding of higher chemical affinity (Fig. 2.3).

The sequence of amino acids is the key to folding of proteins; a wrong sequence will result in unsuccessful folding. A sequence that successfully folds has an extremely well-designed arrangement of amino acid residues which allows formation of many intramolecular hydrogen bonds, positioning of the hydrophobic residues to face inside, and positioning of hydrophilic residues to face outside, all contributing to the stabilization of the entire structure. The forces involved in the stabilization of protein folding are complicated because there are twenty different amino acids, as opposed to four in the case of nucleic acids. However, each molecule chooses one conformation repeatedly, based on the principle of minimization of free energy.

Once folded as described above, proteins associate with each other to form larger components[1]. An example of such complexes is human hemoglobin, a tetramer consisting of four protein subunits. In the case of hemoglobin, being a tetramer makes the four oxygen-binding sites work in coordination, resulting in high oxygen-carrying efficiency. It is also often the case that the complexes found in cells are not entirely proteins but rather are combinations of nucleic acids and proteins. A ribosome, which decodes the information in nucleic acids and translates it into amino acid sequence of proteins, is a complex consisting of five RNAs (ribonucleic acids) and about 80 proteins. Association reactions of such intricate molecular complexes are also governed by the principle of minimization of free energy.

[1] The association of protein molecules is caused by weak forces derived from electrostatic interactions other than covalent binding.

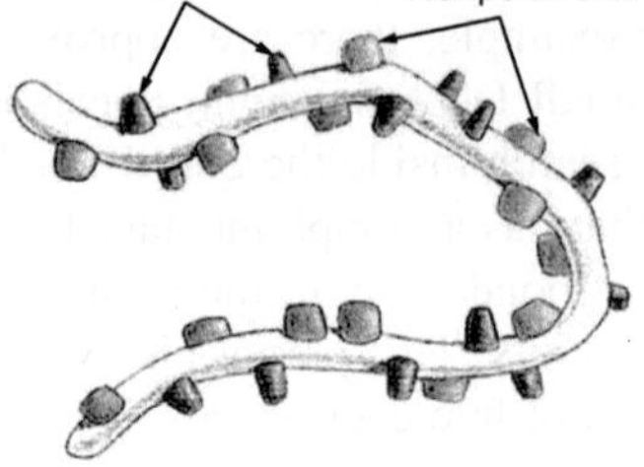

(a) Unfolded polypeptide

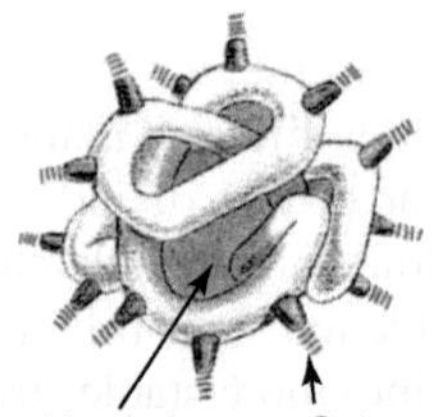

(b) A structure folded in water

Fig. 2.3 Self-folding of proteins (The Cell 3rd Edition ©1994 Garland Science, 1994)

The successful protein folding and assembly of proteins into complex structures all depend on the amino acid sequence. Given a protein consisting of a sequence of one hundred amino acids, the size of the space of possible sequence combinations is in the order of 20^{100}. Selecting a random amino sequence from this combinatorial space would only yield something like sludge with no function. Only a small fraction of possible sequences carry biological significance. Selecting those sequences is beyond the capability of a selection process based on thermodynamics (minimization of free energy). There must be another principle behind the selection; in the biological world this is genetic selection, through which various sequences are tried and tested at the cellular level and the individual level, only those that function well surviving.

2.4 Central Dogma

Central dogma (Fig. 2.4) is a classical concept in molecular biology which states that the genetic information stored in DNA is relayed in the following order:

DNA → (replication) → DNA → (transcription) → RNA → (translation) → protein

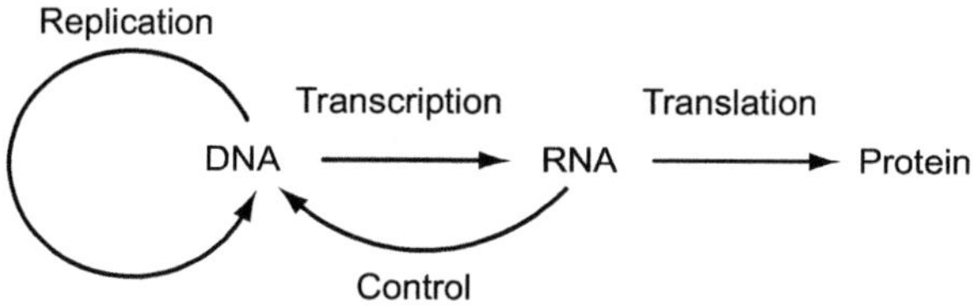

Fig. 2.4 Central dogma

Nowadays, it is not considered dogma anymore, because other information paths have been discovered, but the above is the primary synthesis path for proteins.

In this section, we briefly describe the features of the reactions involved in replication, transcription and translation. *Replication* of DNA from DNA happens during cell division, in order to provide a copy of genetic information to the newly generated cell. An enzyme called DNA polymerase unwinds the DNA double helix so that the bases in the strands are exposed. DNA polymerase then replicates the original sequences by matching complementary nucleotides one by one. *Transcription* of DNA to RNA initiates the protein synthesis, which happens inside nuclei. An enzyme called RNA polymerase copies only the segments of DNA to RNA that carry information on amino acid sequences for proteins (called exons) through a process similar to DNA replication. Such RNAs are called messenger RNAs (mRNA). Then, mRNA is transported out of the nucleus and taken in by a complex molecular machine called a ribosome. The ribosome then synthesizes a chain of amino acids by matching up three bases in the RNA to one amino acid (this match-up is regarded as decoding by the genetic code table). This process is called *translation*. Chains of amino acids synthesized in this way will gain their functions when they are folded.

The genetic information of DNA (instructions for synthesis of proteins) does not just go through the processes of transcription and translation. The transcription process is controlled by complex feedback and feedforward chemical reactions.

2.5 Biological Development: Assembly at the Level of Cells

Cells multiply by division. A monocellular organism consisting of a single cell, such as E-coli and yeast, simply creates exact copies of itself. As evolution proceeded, however, cells came to divide into different cells, and those cells did not separate from each other but rather formed a multicellular organization. In the multicellular organisms, although the genetic information carried by each cell is the same, the way that information is used ("expressed") is different in different cells, resulting in differentiation of cells to form nerves, muscles, blood cells, digestive organs, and so on.

All animals and plants on earth are multicellular organisms, and all are created by cell division starting from a single cell, a zygote. One cell splits into two, and then four, and gradually the number of cells keeps increasing to as many as three trillion in the case of human babies, which then increases further up to 60 trillion as they grow to be adults. Some simpler animals, for example nematodes (C. elegans) which have only about a thousand cells in the whole body, have their *cell lineage trees,* that identify what cells divided in what ways to form each part of the body, fully mapped by researchers (Fig. 2.5). These trees show that the developmental fates of cells whose characteristics diverge after each division are precisely programmed.

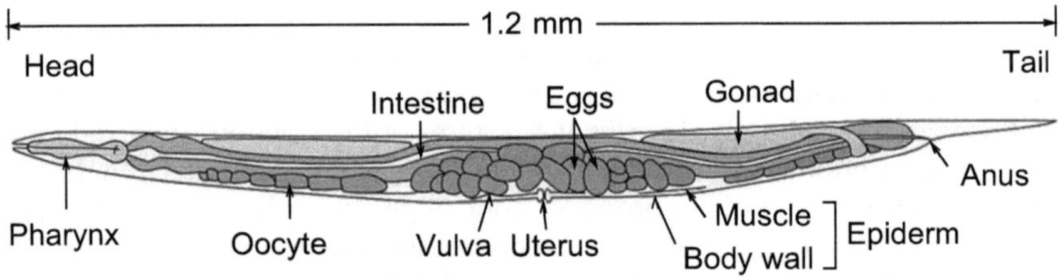

(a) Nematode (C. elegans) (The Cell 3rd Edition ©1994 Garland Science).

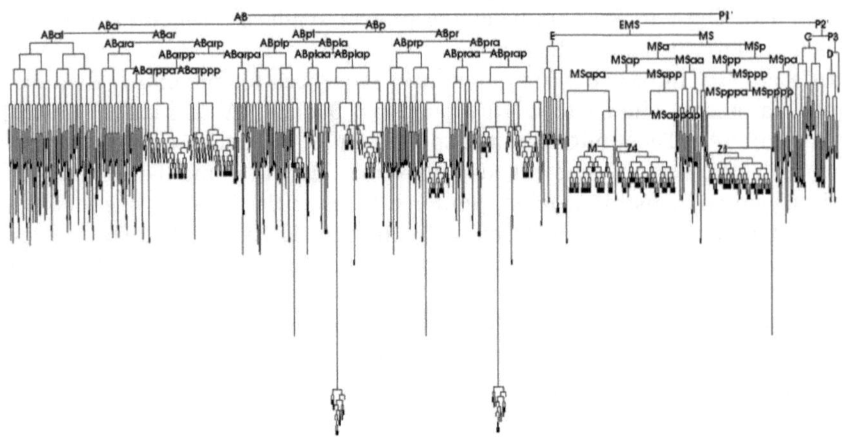

(b) Cell lineage (http://upload.wikimedia.org/wikipedia/commons/3/37/Complete_cell_lineage_of_C_elegans.png)

Fig. 2.5 Cell lineage of C. elegans

For animals with a larger number of cells, such as frogs, mice and humans, lineage trees for each cell usually cannot be mapped. Instead, the development line of sheet-like layers of cells, called germ layers, has been studied. During the first few cleavages, cells are still uniform (totipotent), but then they gradually differentiate and form three different layers called ectoderm, mesoderm and endoderm. The ectoderm forms the brain, nerves and the epidermis (skins), the mesoderm forms the skeleton, the muscle and blood, and the endoderm forms the epithelium of respiratory organs and digestive organs. Lineage trees for such groups of cells (germ layer lineages) have been mapped.

These developmental processes are controlled by interactions between germ layers. In order for germ layers to interact, there is a need for substances that allow cells at a distance to influence each other. Indeed, various mechanisms have been discovered in which proteins and small molecules permeate out of cells and diffuse to give signals to remote cells (Fig. 2.6).

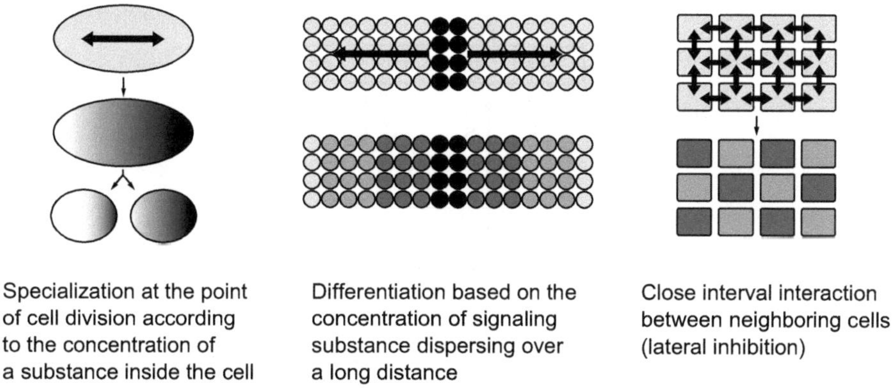

Fig. 2.6 Differentiation strategy based on substance diffusion (The Cell 3rd Edition ©1994 Garland Science, 1994)

Below, we consider a mathematical model of cellular interaction based on diffusion, proposed by Lewis Wolpert in 1969 [4]. In this model, we posit a hypothetical substance called *morphogen*, rather than a specific chemical substance, which carries information about "location."

As an example, let us consider a case where cells of red, blue and white colors form a tricolor pattern like the French flag. (The colors can be taken to represent body parts such as head, abdomen and legs so that this becomes a model of body formation.) In order to generate such a pattern, each cell must know its location relative to the whole in order to determine its color. Now, we assume that the hypothetical morphogen L is generated and released from a cell located at the left edge, and that L is decomposed by a cell located at the right edge, so that there is a gradient of concentration of L across the body. Then, by introducing two thresholds of L concentration for determining which color each cell should exhibit, the tricolor flag pattern will be formed (Fig. 2.7).

This tricolor flag model also gives an explanation of the process of self-repair: assume that the flag has been partially burned, and one third of the flag from the left has been lost. Then, the cell which now is located at the left edge of the remaining part will recognize that it has nothing on its left anymore and start generating morphogen L. Further, a steeper gradient of concentration can now be established. Finally, after the cells go through the differentiation process using the same two thresholds, the tricolor flag is reproduced, albeit two thirds the original size. This explanation uses the hypothetical morphogen, but if one thinks of it as a kind of molecule, it corresponds to the regulation mechanism of gene expression.

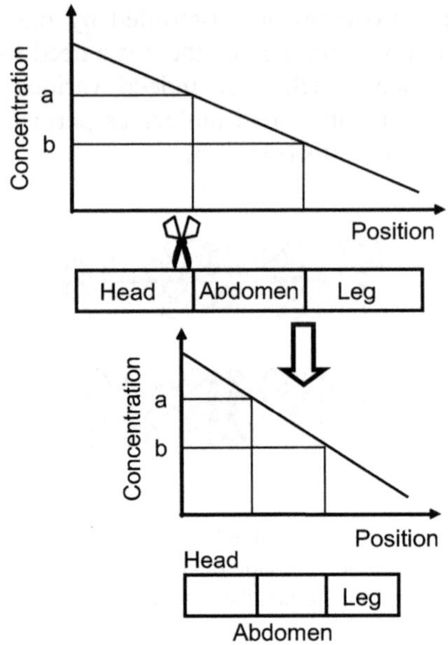

Fig. 2.7 Tricolor flag problem

Actually, many different molecules have been discovered that function in the same way as morphogen in the development process of animals and plants.

Readers may think that such determination of developmental fates based on the concentration of only one type of morphogen and a few thresholds would be too sensitive to small fluctuations. For example, if there is some problem that causes morphogen to leak out while diffusing, the concentration gradient pattern will be disrupted, which will result in formation of a distorted pattern. To avoid such a situation, real biological systems have *cascading* mechanisms that use several different types of morphogen triggered in a time sequence (Fig. 2.8). These act as mechanisms which limit the dependence of cell differentiation on concentration fluctuations.

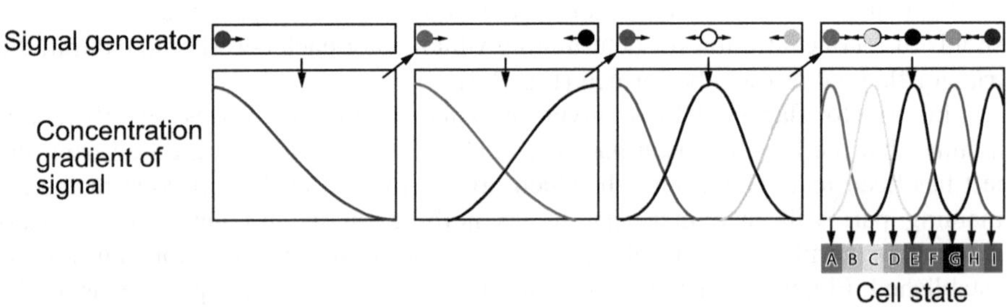

Fig. 2.8 Cascading of morphogen (The Cell 3rd Edition ©1994 Garland Science)

There is also a mechanism that allows generation of various patterns without increasing the number of types of morphogen. It is the reaction-diffusion model, proposed in 1952 by Alan Turing, who is well known for his model of computation. This model assumes two kinds of morphogen, activator and inhibitor, which not only diffuse but also react with each other. High levels of activator will induce generation of more activator, while high levels of inhibitor will suppress generation of activator. The model also assumes that inhibitor diffuses faster than activator (See Section 4.2.3) [5].

Solving partial differential equations obtained from the above assumptions yields various stripe patterns with different widths depending on the coefficients for reaction and diffusion. Fig. 2.9 shows a comparison between a two dimensional numerical simulation by Shigeru Kondo and the patterns on a skin of emperor angelfish [6]. There clearly is amazing similarity between them. The reaction-diffusion model generates various striped and dotted patterns without any special boundary conditions on morphogen as in the tricolor flag case where morphogen is generated at the left end and decomposed at the right end, and therefore the model is thought to explain colors of animal fur, patterns of shells, and arrangements of plant leaves.

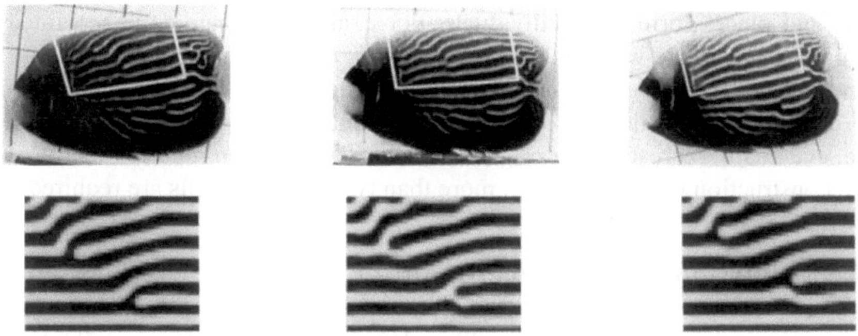

Fig. 2.9 Stripe patterns on the body surface of emperor angelfish and 2-dimensional Turing model (Kondo S, Asai R (1995) A reaction-diffusion wave on the skin of the marine angelfish Pomacanthus, Nat 376:765-768 ©1995 NPG)

2.6 Biological Self-repair

During the process of development, multicellular organisms not only organize and build up their structure, but also have the capability to reconstruct their structure. In this section, we focus on biological self-repair.

Monocellular organisms can create a new individual by simple cell division. In most multicellular organisms, however, it takes two individuals (parents) to reproduce, and moreover, the process by which the reproduced individual develops from a zygote to an adult is very long. The life of an individual is longer than that of a cell, which means that cells in the body must be replaced as they die. This requirement led to the evolution of self-repair capability. In this sense, development and self-repair are the two aspects of the same phenomenon in multicellular organisms.

There are many variations in biological self-repair capabilities. First of all, that of animals is fundamentally different from that of plants. In fact, for plants it is difficult to define the notion of individual clearly, and therefore, in the case of plant regeneration it is difficult to distinguish between self-repair and generation of a new individual[2].

Below, we focus on self-repair of animals. Animal self-repair is categorized into the following types.

2.6.1 *Reconstruction*

The process in which the cells of an organism which has been completely torn apart into pieces form new individuals. For example, sponge cells separated by sifting through gauze will gradually gather to form aggregations and then reattach with each other to become multicellular organisms because of the cells' mobility and adhesion among themselves. They then continue to proliferate and differentiate, eventually recreating the original structure. It has been reported that in order for reconstruction to be successful, more than two thousand cells are required.

2.6.2 *Physiological Regeneration*

The continuous process of loss of cells and replenishment with new cells. For example, a number of the cells at the tip of intestinal villi die every day and are discarded inside the intestine, while the loss is filled by cells generated by division of cells underneath. Similarly, cornified epidermis cells on the surface of the body are removed naturally, and the loss is supplemented by the shifting of new cells to the basal layer of the epidermis. Physiological regeneration processes like these are achieved by cell generation by stem cells, which retain cell division capability for a long time.

---

[2] Self-repair of animals requires movement of cells, whereas plants repair themselves by replicating their cells at the place. Plant cells are surrounded by cell walls and in principle are incapable of moving from the location where they were born. Cell division starts with formation of a new wall inside the cell, and afterward there is expansion beyond the space of the original cell.

2.6.3 Compensatory Regeneration

When parts of some organs are removed or damaged, that part is regenerated by cell proliferation. It is well known that the liver has high regeneration capability. In the case of mice, the liver will recover its original state in 24 hours after being partly removed. In this particular regeneration, it is not the case that newly generated undifferentiated cells specialize to be liver cells, but rather that liver cells reproduce themselves directly to make up for the loss.

2.6.4 Wound Healing

The process of repairing relatively small wounds of the body. The healing process of a little scratch on the epidermis is an example. This is the smallest self-repair process, but still it involves cell movements, adhesion and differentiation.

2.6.5 True Regeneration

This falls into the following two categories:

Morpholaxis

The typical example is an entire planarian that regenerates from a small piece of body. A cell cluster remaining from an organism regenerates the structure of the entire organism by redifferentiating existing cells using its information on all the parts of the original organism (corresponding to the location information of the tricolor flag pattern discussed in the previous section). After the whole structure is regenerated, cells multiply by division so that the new organism grows to the size of the original.

Epimorphosis

When a leg or a tail of a gecko is amputated, undifferentiated cells proliferate at the stump to form a blastema. This blastema, as it grows, differentiates and develops structure, following processes similar to biological development.

Fig. 2.10 illustrates experiments transplanting planaria body parts into other planaria and cockroach legs into other cockroaches. When the head segments of planaria were transplanted either in front of or to the rear of the pharynx (throat) of other planaria, only in the second case was another pharynx formed. The body was mapped with 15 coordinate points lengthwise. When a head segment of one, from point 1 to 2, was transplanted into point 5 of another, the segment regenerated only the part corresponding to 2 to 5. On the other hand, when the same segment was transplanted into point 12, it regenerated the part corresponding to 2 to 11. A similar phenomenon was observed in experiments on transplanting cockroach legs.

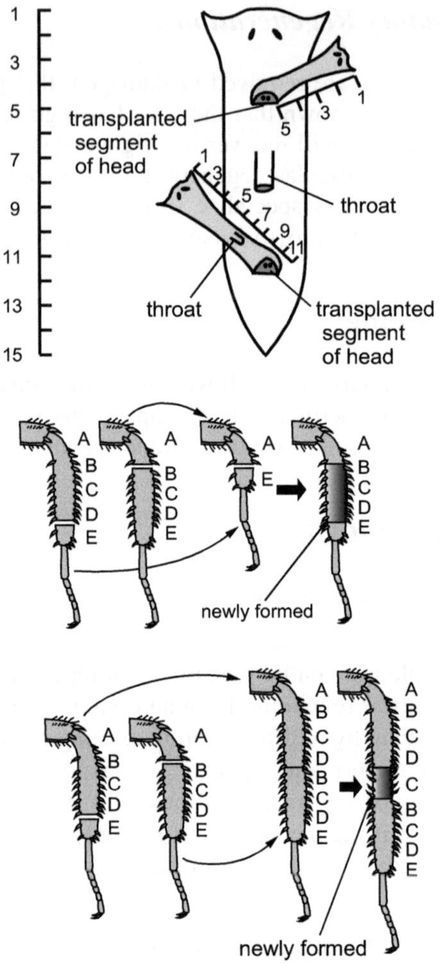

Fig. 2.10 Regeneration experiments on planaria and cockroaches (The Cell 3rd Edition ©1994 Garland Science)

2.7 Self-Organization of a Group of Individuals

The last section of this chapter presents various examples of self-organization in the highest level of our hierarchy, individual biological organisms. In this section, we consider cellular slime molds, social insects such as ants and bees, schools of fish and flocks of migrating birds.

2.7.1 *Cellular Slime Molds*

Cellular slime molds are unusual creatures which sometimes live individually as monocellular organisms but at other times form multicellular bodies. As monocellular organisms, they are amoebae that move individually. When environmental conditions deteriorate, however, for example if food is running out, many amoebae gather to form a slug-like entity, which then moves as a whole. The system of interaction between slime molds in the cellular phase that causes this to happen involves reaction-diffusion of a substance called cyclic AMP. The slug-like organism eventually produces a fruiting body as in fungi, and releases spores from the tip. Spores which land in a good environment will hatch into amoebae, which multiply and eventually form the next slug-like organism.

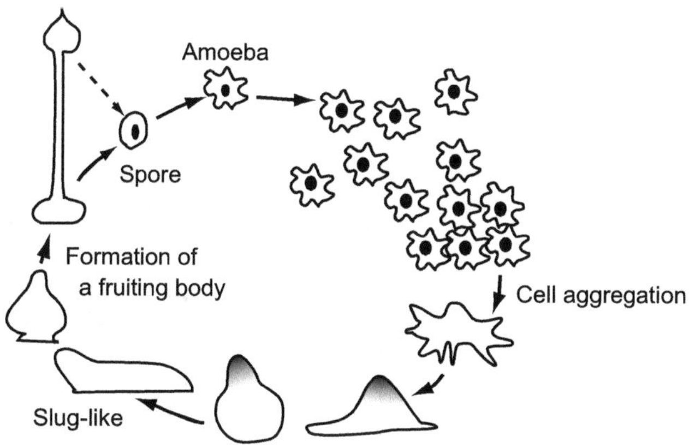

Fig. 2.11 The life cycle of cellular slime molds

2.7.2 *Social Insects*

Ants and bees are sometimes called social insects. In the case of bees, queens, workers, and male drones all have different morphologies, and they form a hierarchy with a single queen bee at the top. Worker bees, which are female, are in charge of almost all the daily maintenance, while all bees in a colony are produced by the queen. The only contribution of males is to provide genes. In recent years, much ethological research on insect has been directed at trying to understand their different behavior patterns as contributing to self-organization. So far, attempts have been made to explain apparently complex phenomena, such as nest making and “dancing” of bees and the movement of ants from the nest to food by the shortest route, as the result of following relatively simple rules [7]. There are engineering-oriented studies of social insects which are being applied to solve large scale search problems, called ant colony algorithms.

2.7.3 Herds of Animals

Animals gather to form herds to make survival more likely, because this affords protection against predators, makes breeding easier, etc. Animal herd behavior takes various forms, including schools of fish staying close together and swimming in the same direction, flocks of migrating birds in formation, and a troop of monkeys who stay in groups although the behavior of each individual is different. The dynamics dominating schools of fish and flocks of birds have been well studied, and various mathematical models have been suggested. Fig. 2.12 shows a model, which creates various behavioral patterns by varying a single parameter [7].

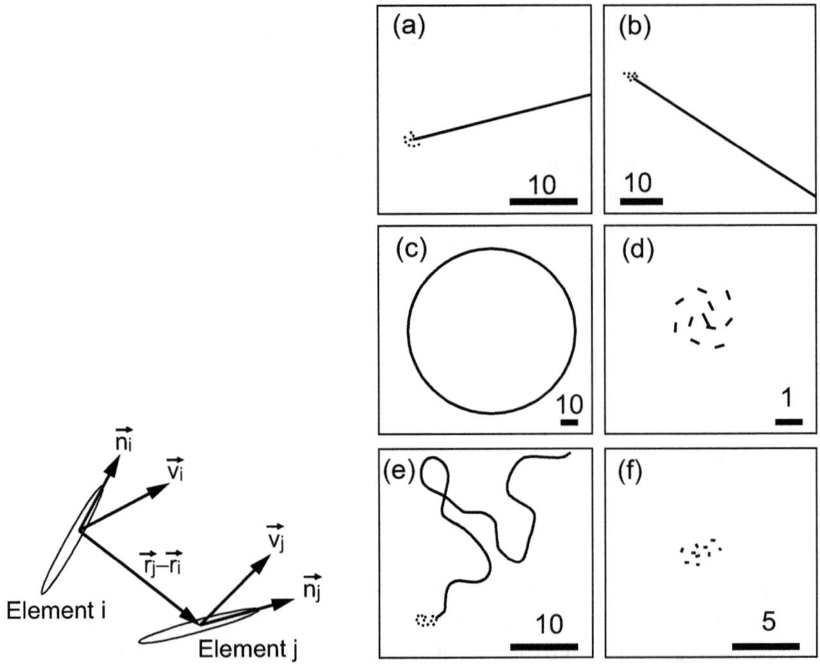

Fig. 2.12 Herd behavior model. By positing simple interaction among individuals, various behavior patterns can be generated, such as straight line movement (marching), zig-zagging, moving in a circle, and random movement. (Shimoyama N et al (1996) Collective Motion in a System of Mobile Elements, Phys Rev Lett 76(20):3870-3873 ©1996 APS)

References

[1] Alberts, B., et al.: Molecular Biology of The Cell, 3rd edn., Garland Sci. (1994)
[2] Watson, J.D., et al.: Molecular Biology of The Gene, 5th edn. Cold Spring Harbor Lab Press (2003)
[3] Ito, M.: Construction of Decentralized Autonomous Systems. J. Soc. Instrum. Control Eng. (Keisoku to Seigyo) 29(10) (1990) (in Japanese)
[4] Wolpert, L.: Positional Information and the Spatial Pattern of Cellular Differentiation. J. Theor. Biol. 5, 1–47 (1969)
[5] Turing, A.M.: The Chemical Basis of Morphogenesis. Philos. Trans. Royal Soc. Lond. B 237(641), 37–72 (1952)
[6] Kondo, S., Asai, R.: A reaction-diffusion wave on the skin of the marine angelfish Pomacanthus. Nature 376, 765–768 (1995)
[7] Shimoyama, N., et al.: Collective Motion in a System of Mobile Elements. Phys. Rev. Lett. 76(20), 3870–3873 (1996)

Chapter 3
History of Self-Organizing Machines

Abstract. In this chapter, we give an overview of research related to self-organizing machines that was inspired by biological self-organization. There were various motivations for this kind of research. Von Neumann's work, which we discuss first, was motivated by the purely mathematical question of whether self-reproduction is logically possible or not. Next, we discuss the work by Penrose, who considered the above as a question of actual objects, to be solved with experiments. We then go on to consider some of the studies that followed upon these two. A basso continuo underlying all these studies is a desire to understand the mechanisms behind the biological organisms discussed in the previous chapter. In order to understand these mechanisms, instead of studying biological organisms directly, they focused on some aspects of biological organisms of interest and modeled them. This is called a *constructive approach*[1]. If biological organisms can be understood through a constructive approach, machines can be built according to the derived model.

3.1 Work by von Neumann

The first appearance of the notion of self-organizing machines, or machines capable of modifying themselves, was in the work by John von Neumann, dating back to around 1950 [1]. He proposed a basic framework by which a machine can assemble itself from parts using a *description* of its own design or assembly procedure embedded in the machine. This separation of the system's structural description and its physical embodiment, also present in the genotype and phenotype of biological organisms, is essential for a system to have the capability for evolution. In this sense, von Neumann's work still has a definitive influence on self-organizing machine research [2].

---

[1] When trying to understand biological organisms, it is often the case that they are difficult to analyze because they consist of a very large number of components assembled in a very complex way. Molecular biology offers a very powerful method of tracing the origin of biological properties to those of substances, that is, genes and proteins. On the other hand, it is still difficult to trace the origins of functions that themselves are not yet understood, such as regulation of the size of the body, to the genes. This is the rationale for building models of simplified biological properties.

S. Murata and H. Kurokawa: Self-Organizing Robots, STAR 77, pp. 37–57.
springerlink.com

One group of research that originated from von Neumann's work is the theory of cellular automata, which later has been applied to areas such as fluid dynamics and relatively abstract models of living organisms such as artificial life. Another current of research went to the self-organization of real physical systems such as Penrose's blocks [3, 4]. It is the last line of research, combined with the development of electronics and robotics that followed, that gave birth to the module based self-organizing machines described in this book (More history is found in Chapter 6. See Fig. 6.5).

3.1.1 Von Neumann's Two Questions

John von Neumann was a mathematician born in Hungary who worked in the U.S.A. He made outstanding contributions in various areas of applied mathematics including quantum mechanics, establishing game theory and reliability theory. He is renowned for his involvement in the development of EDVAC, one of the very first electronic computers. Modern computers have what is known as the von Neumann architecture (with internally stored programs). He was also involved in the development of the atomic bomb during World War II and is known for his hawkish opinions on politics.

While working on the development of electronic computers, von Neumann tried to establish a general theory for computers. In particular, he noticed the similarities and differences between computers and living organisms in nature and tried to categorize them systematically. In his last years, he was engaged in *the theory of self- reproducing automata*[2] [1].

According to Arthur Burks, who edited von Neumann's posthumous writings, "*Von Neumann compared the best computers that could be built at the time with the most intelligent natural organisms and concluded that there were three fundamental factors limiting the engineer's ability to build really powerful computers: the size of the available components, the reliability of these components, and a lack of a theory of the logical organization of complicated systems of computing elements... He felt that there are substantially different principles involved in systems of great complexity, and searched for these principles in the aspects of self-reproduction which clearly depend on complexity. Because of the close relation of self-reproduction to self-repair, results of self-reproduction research should also help solve reliability problems.* [1]"

In summary, the two fundamental questions that von Neumann posed were:

1. How can reliable systems be constructed from unreliable components?
2. What kind of logical organization is sufficient for an automaton to be able to reproduce itself?

The most intelligent biological structure in nature is a brain, which is comprised of neurons. If we aim to build a system equivalent to a brain, we need artificial components equivalent to neurons.

---

[2] An automaton is a system whose inputs, internal states and outputs are defined mathematically. For more details refer to textbooks on computer science or formal language theory.

Prior to the above argument by von Neumann, Warren McCulloch and Walter Pitts suggested that neuron structure can be represented using logic circuit components [5]. Inspired by this work, von Neumann created a theory for constructing intelligent systems from artificial components such as vacuum tubes and mechanical relays. The difference between von Neumann's theory and the pure mathematics (formal logic) is that he took into consideration issues such as delays of input/output and errors occurring with very low but non-zero possibility. The biggest challenge in building complex artificial systems is to detect malfunctions in those components without fail and to replace or repair those broken components promptly.

Related to the first question above, consider transmitting the value one million in base ten; the simplest way is to send each of seven digits,"1", "0", "0", "0", "0", "0", and "0". However, this is not a good transmission method because of the possibility that wrong digits may be received due to some error. Instead, sending a pulse a million times to be counted on the receiver side would offer a better probability of successful communication. This line of discussion led to the establishment of reliability engineering in later years.

As for the second question, the key is the complexity of automata. The difference between automata in nature and artificial ones lies in their complexity. What artificial automata (machines) produce is usually simpler than the machines themselves, and hence the complexity decreases in the process. For instance, if you compare a machine that produces screws with the screws that it produces, the machine itself is much more complex than the screws. On the other hand, living organisms in nature can reproduce exact copies of themselves, and therefore the complexity does not decrease in the process. In fact, through the long evolution process, complexity is gradually increasing rather than decreasing. Is there an explanation for this phenomenon?

3.1.2 Von Neumann's Self-reproducing Automata

In order to investigate this question rigorously, von Neumann studied Turing's universal automata [6], whose inputs, outputs and internal states take only two values, 0 and 1, which simplifies the discussion.

A universal automaton is an automaton that can simulate the behavior of any other automata. To construct a universal automaton, one needs a description (specification) of the automaton to be simulated and the instructions (input) that are given to that automaton. There are automatons whose input-output characteristics can be expressed as a sequence of binary numbers without difficulty. For instance, if the automaton is a simple logic gate, a single sequence that lists the contents of its truth table will completely describe it.

Now, let $\phi(X)$ be a binary sequence describing a certain automaton X. We call this a "description." Also assume the existence of a universal automation A (universal constructing machine), which can construct X while gradually eating up the description $\phi(X)$. The construction method of such a universal automaton A is the core of von Neumann's theory, which we explain later.

Von Neumann claimed that if there actually is a universal automaton, a system could be made that would reproduce itself with increased total complexity (Fig. 3.1). However, we need more tools to accomplish that. By feeding ϕ(A) to the automaton A, A will produce a copy of A itself, but without increase in total complexity. In fact, since this process consumes ϕ(A), the resulting complexity is reduced.

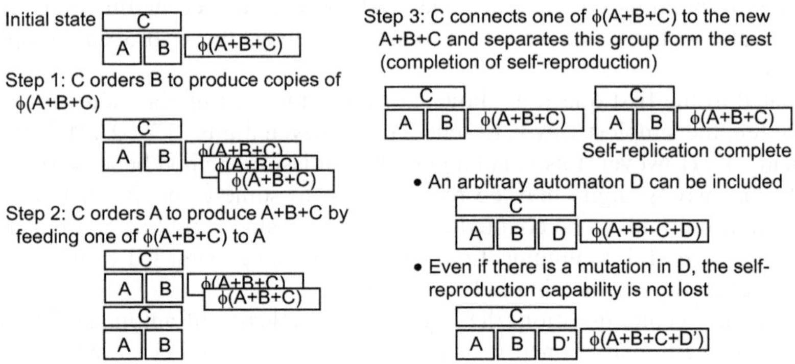

Fig. 3.1 Von Neumann's self-reproducing system

So, as a next step, we provide another automaton B which, when given description ϕ(X), produces two copies of ϕ(X) without consuming the original ϕ(X) (description copying machine). B is only capable of producing copies of descriptions but not B itself, and it does not increase the total complexity. Now, we add an automaton C, an appropriate control device for the automaton A+B. The automaton C gives orders to and functions as a control device of both A and B according to the following sequence.

Starting from a state where both the automaton (A+B+C) and its description ϕ(A+B+C) exist (Fig. 3.1), first, C orders B to produce two copies of the description ϕ(A+B+C). Then, C orders A to produce (A+B+C) using one of the copies of ϕ(A+B+C). Finally, C attaches one of the two remaining descriptions ϕ(A+B+C) to the newly produced (A+B+C). The original pair of the automaton and its description, i.e. (A+B+C) and ϕ(A+B+C), has produced a copy of the pair, with no decrease in complexity.

One important feature of this system is that A, B, and C each do not have self-reproduction capability, but when those three are combined, this capability is achieved. In other words, a self-reproducing system can be decomposed into components each of which does not have self-reproduction capability.

Moreover, this model also gives an account of the process of evolution. A specific automaton D is introduced, and if the description ϕ(A+B+C+D) is supplied, the self-reproducing system can be extended to include D. This system not only reproduces itself but also produces a byproduct. Now assume that some mutation happens to such a system. If the mutation happens in components indispensible for self-reproduction, that is, in either A, B or C, then the system loses its ability to self-reproduce (lethal mutations), but if it happens in D, since D does not contribute to reproduction, the mutated version of D will be reproduced.

In this way, individuals with various mutations are produced, and those that are better suited to the surrounding environment will succeed in producing surviving descendants. That is, the mechanism of evolution based on survival of the fittest, natural selection, is set into operation. This model demonstrates how artificial entities can increase in complexity.

3.1.3 Universal Automata: The Kinetic Model

In the previous section, we postponed explaining the details of how to construct an automaton that is capable of producing an automaton from a description. In fact, this is a difficult problem.

Von Neumann's first solution was called a *kinetic model*. This model deals with geometric and kinematic issues such as the movement, contact, positioning, fusing and cutting of components, while ignoring the physical and chemical issues such as force and energy. A three-dimensional space is envisioned in which various parts are floating around. A self-reproducing automaton is made from these parts. We want this automaton to be capable of picking up the parts around it and assembling them into a copy of itself. As explained above, it is easier to let the automaton refer to a description tape, a binary sequence specifying its own structure, than to let the automaton examine and determine its own structure. In von Neumann's draft, description tapes made of connected rods were introduced as shown in Fig. 3.2.

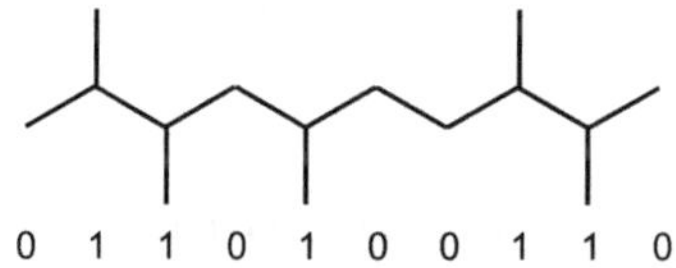

Fig. 3.2 Example of a description using rigid members. A joint with a side branch represents 1, while one without represents 0.

Von Neumann used eight kinds of parts in his kinetic model, all of which are expressed as straight lines (line segments). Four of the eight kinds are information processing organs (parts). A *stimulus organ* receives and transmits stimuli; and at the same time performs logical addition "p or q." A *coincidence organ* realizes the truth-function "p and q." An *inhibitory organ* realizes the truth-function "p and not q." A *stimulus producer* (pulse generator) serves as a source of stimuli. The remaining four kinds are in charge of functions other than information processing.

A *rigid member* is a structural part with which a framework for automata and description tapes can be constructed, and it does not carry any stimuli. Two parts are connected using a *fusing organ*. In order to connect member *a* to another member *b*, the output end of the fusing organ should be placed at the contact point of members *a* and *b*. A stimulus provided at the input end of the fusing organ at time *t* causes members *a* and *b* to be welded at time t+1. On the other hand, a *cutting organ* breaks a connection when stimulated. A *muscle* is capable of motion, and will contract to length zero at time t+1 after being stimulated at time *t*, while maintaining all its connections.

An automaton constructs another automaton as follows: the parent automaton floats in a sea containing an infinite supply of parts. The parent automaton's memory (the description tape) contains a description of a child automaton. The parent automaton picks up necessary parts and places them appropriately, following the instructions given in the description.

The parent automaton has two protruding stimulus elements. When a part touches them, they stimulate that part and detect the part's response to determine what the part is. For example, if the part is a stimulus organ, a stimulus is transmitted, whereas if it is a muscle, it contracts.

After thinking through this much, however, von Neumann became stuck. In order to make this model more concrete, he had to specify in detail the spatial placements, connections, characteristics, and movements of the eight kinds of parts, which clearly is a very complicated task. Furthermore, even if all this were done, would the model with its idealized, abstract parts that have no mechanical or chemical aspects at all be of any value? This may have been his thinking when he decided to give up on developing this kinetic model further.

The authors still find this depiction of self-reproducing automata deeply inspiring, because a living organism seen as a molecular machine is indeed a realization of such a universal automaton, and von Neumann's model extracts its essence. It is a vivid image of what an artificial self-reproducing system must be if it is to exist at all.

3.1.4 Universal Automata: The Cellular Model

Von Neumann next created the cellular model. This universal automaton is constructed as an assembly of many cells in various states, arranged in a certain spatial pattern. A system can be defined as an assembly of differentiated cells in a lattice space filled with undifferentiated cells. By introducing this framework, the difficult problem for the kinetic model of how to secure a supply of parts is resolved.

In this model, cells fill up the spaces of a lattice like a sheet of graph paper. Each cell is in one of 29 states and the state determines the cell's function (Fig. 3.3). Each cell's function is determined by the cell's state and its inputs from its neighbors. An ordinary transmission cell, a cell in one of the ordinary transmission states, sends a pulse in the direction indicated by the arrow within one time unit. When this cell is sending a pulse, it is marked with a dot to identify its state.

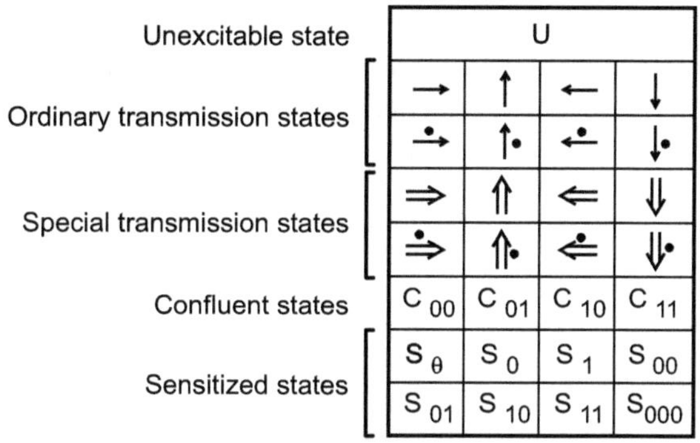

Fig. 3.3 Cellular model table of 29 states that each cell can assume (Neumann Jv, (Burks A Ed.) (1966) Theory of Self-Reproducing Automata, Univ Illinois Press)

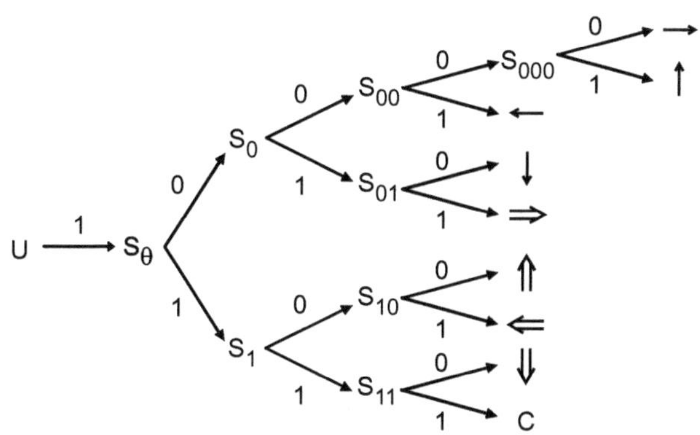

Fig. 3.4 Transition rules (Neumann Jv, (Burks A Ed.) (1966) Theory of Self-Reproducing Automata, Univ Illinois Press)

Since this cell responds to an input pulse from any direction by outputting a pulse in the direction of the arrow, it also works as a logical OR operator. A special transmission cell transmits a pulse as an ordinary transmission cell does, but ordinary transmission cells (→) and confluent cells (C) change back to ground state (unexcitable) cells (U) when they receive a pulse from a special transmission cell (⇒). A confluent cell transmits a pulse during two time units, and it is used for branching and logical AND operations. A ground cell receives a pulse sequence (a binary sequence) and changes to one of the transmission states or confluent states via one of the sensitized states, which are states of cells undergoing differentiation. When pulse sequences are received by a cell, its path of differentiation toward a specific type of cell is predetermined by transition rules (Fig. 3.4).

In the actual process of self-reproduction, a pattern of cells in a certain location extends an arm to a different location and creates the same pattern. Specifically, the arm refers to a description (a sequence of cells that records a pulse sequence) to change the state of a cell far away. This process is called the generation of a cell. Next, the arm extending to that cell, which itself is a concatenation of cells in transmission states, is erased. By repeating these steps, the pattern is written (Fig. 3.5).

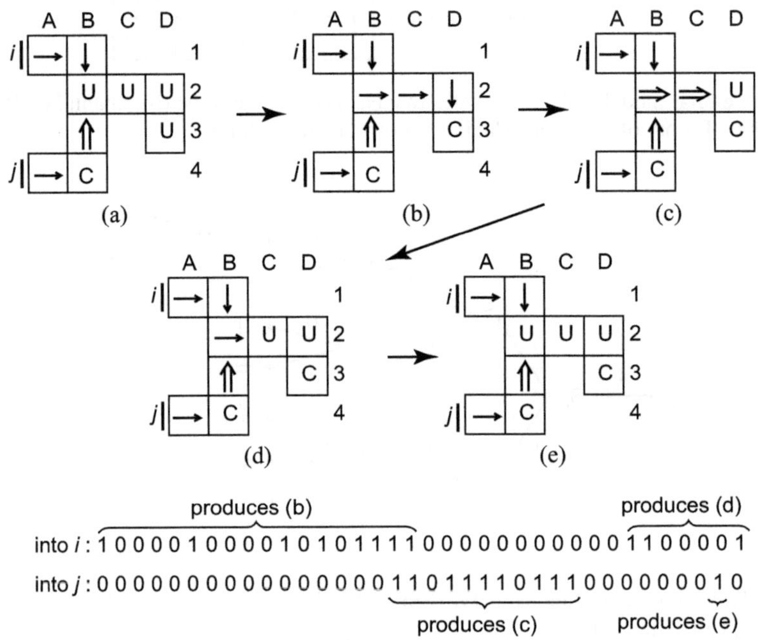

(f) Supplied pulse sequences

Note: "0 arrows" and "1 arrows", $\underset{\rightarrow}{0}$, ↑1, etc., in the original are represented as "arrows" and "double arrows", →, ⇑, etc., here to ensure consistency with other figures.

Fig. 3.5 Process of assembly using pulse descriptions (Neumann Jv, (Burks A Ed.) (1966) Theory of Self-Reproducing Automata, Univ Illinois Press)

The actual system requires control of the arm and circuits for generating various signals, so that as many as 200,000 cells are required to construct the automaton (Fig. 3.6). It is quite amazing that von Neumann invented this automaton using only pen and paper. In recent years, people have confirmed by computer simulation that the automaton described above is indeed capable of reproducing itself.

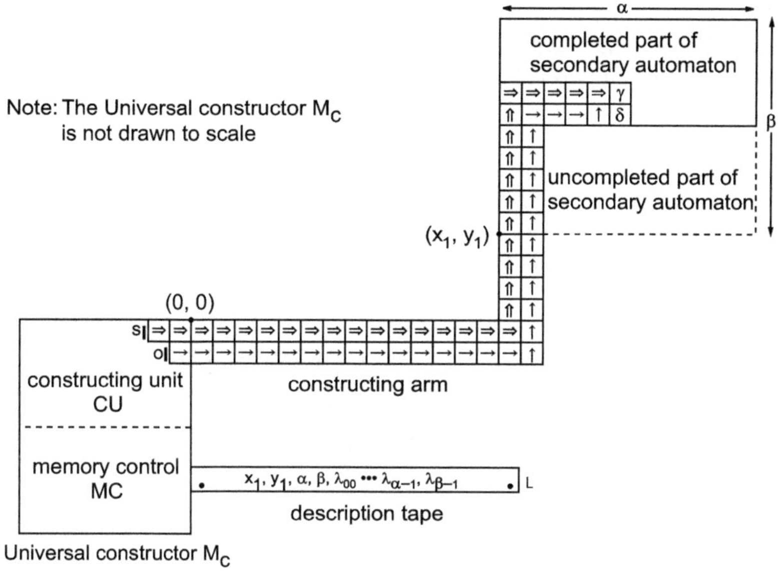

Fig. 3.6 Entire structure of von Neumann's self reproducing automaton (Neumann Jv, (Burks A Ed.) (1966) Theory of Self-Reproducing Automata, Univ Illinois Press)

3.2 Work by Penrose

Lionel S. Penrose was a British psychiatrist and geneticist who also made various contributions in mathematics. One such contribution is his block models of self-reproducing machines which we explain in this section [3, 4]. The models come in varying complexity, from simple to very complex. Here we explain a one-dimensional model which is easy to understand. Here, "self-reproduction" is defined as the generation by a system of new systems which have the same structure and size as the original, when the system is placed in a suitable environment. Further, in order to exclude simple chain reactions we impose the following conditions:

1. A self-reproducing system consists of plural components, and those components must be simpler than the system itself.
2. Different self-reproducing systems with different combinations of components can be made, each of which reproduces itself (the initial self-reproducing system is called a seed).

The first model that Penrose thought of was a system consisting of two kinds of blocks with bilateral symmetry, A and B, lined up in a one-dimensional sheath as in Fig. 3.7. The freedom of movement of each block is limited so that each can only shift or tilt to the right or left. By shaking the sheath right and left, one can make the blocks collide with each other. The blocks have a convexity and concavity such that when two A blocks collide they simply rebound, and similarly for two Bs, but when a tilted block A and a similarly tilted block B collide, a convexity of one and a concavity of the other become joined together. If a joined BA structure is placed as a seed on the track before shaking (ii), only those blocks that happen to be in the order B-A become joined (iv). Conversely, if a joined AB structure is used as a seed (v), then only those in the order A-B become joined (vii). It is easy to see that this system satisfies the above two conditions.

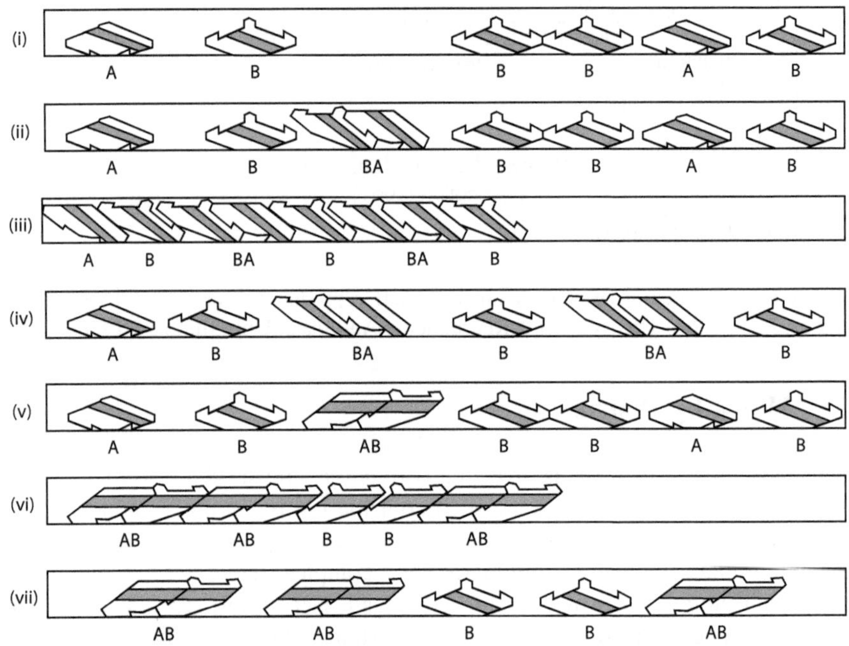

Fig. 3.7 Penrose's blocks (Penrose LS (1958) Mechanics of Self-replication, Ann Human Genetics, 23:59-72)

Penrose claimed that components of self-reproducing systems should have the following five properties:

1. Each component has at least two possible states, non-activated (as an isolated component) and activated (as a part of a system). A non-activated component can be joined only when activated components approach in its vicinity. In other words, when a sheath does not have a seed, and contains only non-activated components, self-reproduction will never happen no matter how much it is shaken.

2. There must be definite boundaries to the activated system. Otherwise, it may grow indefinitely like a crystal[3].
3. The process of self-reproduction must consume some kinetic energy. In the model described above, the kinetic energy supplied by shaking of the sheath is transformed into the energy for linking of activated blocks because the convexity and concavity of the blocks work as a ratchet.
4. The activated state of components is transferred within the system. For example, in the machine shown in Fig. 3.8, activation of the block at the left end (in this case, activation means being capable of independent motion) is transferred to the block on the right end.
5. The movement and interactions of blocks are limited to some extent to make the desired linking more likely. In the one-dimensional case, the sheath provides the necessary limitation, but in general, this is provided by the requirement that something helps the linking, like catalysts.

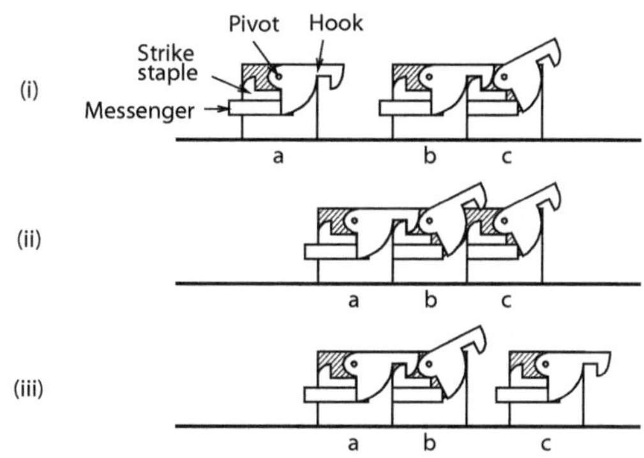

Fig. 3.8 Activation transference mechanism (Penrose LS (1958) Mechanics of Self-replication, Ann Human Genetics, 23:59-72)

Fig. 3.9 shows a one-dimensional self-reproducing system based on the above discussion. Here each block has four layers. The second from the bottom determines the type of the system (whether AB or BA in the case of Fig. 3.7), the top two layers are for linking, and the bottom layer controls the size of a system (so that not more than two blocks are connected). The seed is the two connected blocks in the middle of (a). In (b), the block at the left of the seed is linked to the seed, and in (c), the block at the right of the seed is linked to the seed, at which point all four blocks can be divided in the middle, yielding two seeds as seen in (d). In short, the seed expands by taking in two blocks, and when it becomes twice its original size, it splits into two, generating a copy of itself.

3 Schrödinger said that a living crystal must be aperiodic [7].

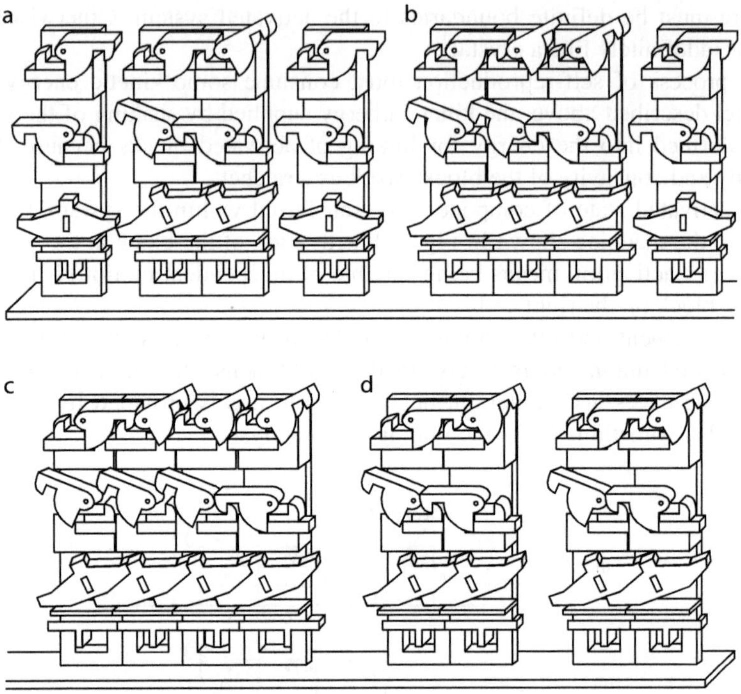

Fig. 3.9 Penrose's self-reproducing block system (Penrose LS (1959) Self-reproducing Machines, Sci Am, 200:105-114)

3.3 Mathematical Models of Self-reproduction

Von Neumann's cellular model has been generalized into the theory of cellular automata and has found applications in areas such as simulations in fluid dynamics. At the same time, research has continued on self-reproducing systems themselves. Since von Neumann's original cellular model was very complicated, people have tried to establish simpler, more comprehensible models.

One study distinct from all others was Langton's work on self-reproducing automata, which we introduce in this section. This work demonstrated the feasibility of self-reproduction using simple, visually understandable cellular automata, and triggered a boom in research into so-called *artificial life*.

We also introduce our own work on a self-reproduction model using graph automata. This model allows more flexible construction of systems by removing the constraint that requires cells to be placed on a two dimensional plane. This constraint has been one of the factors making the cellular model unnecessarily complicated, and this model should thus enable visualization of the essential structure of self-reproducing systems.

3.3.1 Langton's Self-reproducing Loop

Christopher Langton succeeded in greatly reducing the complexity of cellular automata capable of self-reproduction [8]. His model requires only 8 different states, whereas that of von Neumann required 29. Moreover, Langton's model requires only a 10 x 15 cell space. Langton's model simply produces the same pattern as itself, while von Neumann's model was capable of emulating the universal Turing machine. However, the amazing similarity of the behavior of Langton's model to that of living organisms has attracted a lot of attention.

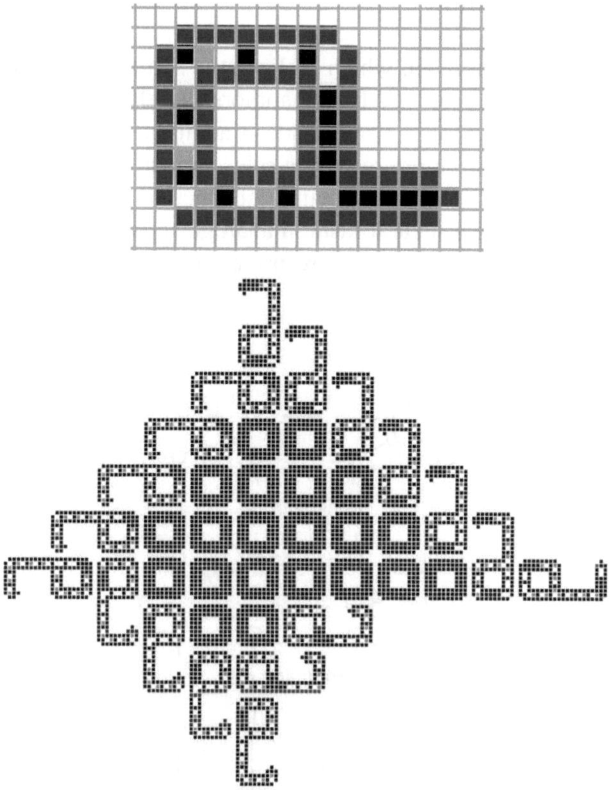

Fig. 3.10 Langton's self-reproducing loop

Langton's loop is a two dimensional cell pattern in the form of a loop, a line of cells containing "genetic information" which is covered by a sheath. The genetic information includes instructions for extending a part of the sheath to grow a kind of arm (pseudopod) and also for bending the arm orthogonally to the left at regular intervals of an appropriate length. The arm continues to extend even after it forms a loop, and forms another empty loop of the same size. Finally, the genetic information of the first loop is transferred into the arm and forms a complete loop, which is then detached from the parent loop. The pattern reproduces itself by repeating these steps (Fig. 3.10).

3.3.2 Graph Automata

A model called graph automata, which realizes the self-reproduction process in a natural fashion, has been proposed by the authors [9, 10]. A graph automaton is a cellular automaton defined on a graph, and, in addition to the state transitions of cellular automata, it has structure rewriting rules so that the structure of the graph itself can be changed. This allows dynamic changes in component numbers and topology to be realized in a natural fashion analogous to self-reproduction of biological organisms, because they are not restricted to a particular lattice space as in the case of cellular automata.

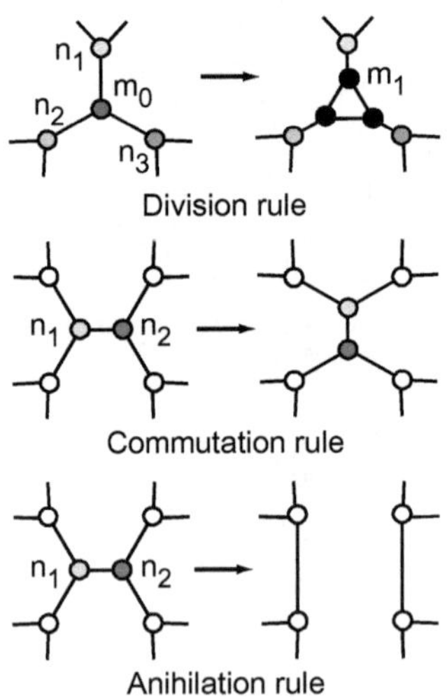

Fig. 3.11 Rules of a graph automaton

The graphs consist of a set of nodes that have internal states and that are connected by links. Here, we assume a planar graph each of whose nodes is adjacent to three other nodes (3-link graph). Three types of structure rewriting rules that we use are described in Fig. 3.11. These rules preserve the property that each node is always adjacent to three other nodes. Each rule is applied at each discrete time step to only those nodes that have specific local states (all nodes are investigated and all applicable rules are applied in the step). Starting from the initial graph, applicable rules in the rule set are applied one after another and the graph is allowed to evolve with time. Since the time evolution is determined by the initial graph and the given rule set, designing a graph automaton is equivalent to determining the initial graph structure and the rule set.

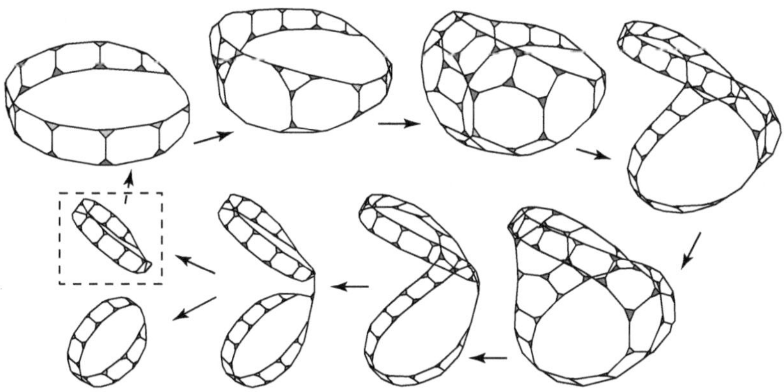

Fig. 3.12 Representation of self-reproducing process using graph automata

As an example of representation using the graph automaton, the self-reproduction of a Turing machine is shown in Fig. 3.12. A Turing machine is the simplest model of computer, and is built here as a ladder-like ring structure. One side of the ladder corresponds to the tape of the Turing machine, while one particular node on the other side corresponds to the reading head of the Turing machine. First, a copy of the entire ladder is produced. Then, the information of the tape is copied on to the new ladder. Finally, the unnecessary links between the two ladders are removed, resulting in two independent ladders. In the case of Turing machines with binary tapes, 20 states and 257 rules are required to reproduce any Turing machine whose tape is of an arbitrary length and contains arbitrary signal sequences.

Thanks to a uniform format for describing rules, it is easy to apply evolutionary algorithms to graph automata. We have been successful in obtaining various self-reproducing patterns by selecting rules with reproduction capability from among randomly generated rules [10].

3.4 Physical Models of Self-reproduction

It may be more appropriate to consider the Penrose block system as a model of self-catalysis in chemical reactions than as a self-reproducing system. In this model, as the units randomly collide with each other, the number of units in particular combinations increases. In this section we consider further developments following Penrose's model [2].

We start with the self-assembly system by Hosokawa which uses magnets. This work analyzes the yields (the ratio of the number of complete assemblies to the number of units initially placed in the container) of artificial self-assembly systems quantitatively for the first time. Next, we consider the mechatronic self-assembly system by Klavins. In contrast to the model by Hosokawa where the only passive magnetic force is used, each of Klavins' units is equipped with a microprocessor, which enables a unit to disconnect from others as desired. In this way, the self-assembly rules can be changed at will and hence flexible programming of the reactions among units is possible. The third example of programmed self-reproduction is Griffith's self-reproducing blocks. These blocks, also equipped with microprocessors, self-reproduce in the same way as Penrose's blocks.

3.4.1 Magnet System by Hosokawa

Kazuo Hosokawa's work on self-assembly of magnetic units is a pioneering quantitative analysis of self-reproducing systems [11]. In the self-assembly process of this model, triangular units with magnets are joined with each other to form hexagons by agitation in a closed two dimensional space (Fig. 3.13).

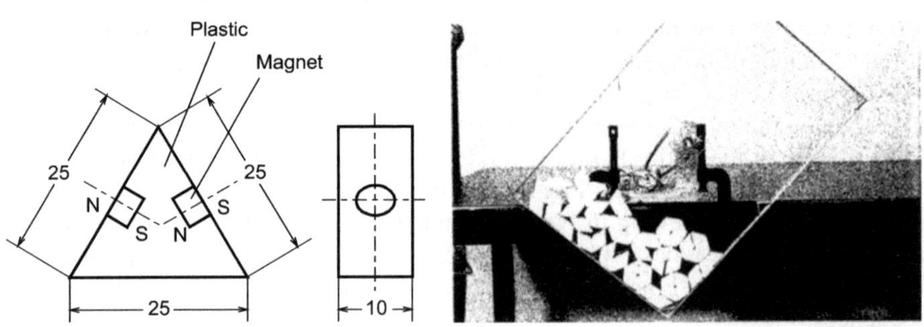

Fig. 3.13 Units with embedded magnets are agitated by rotation in a container (Hosokawa K, et al (1995) Dynamics of Self-Assembling Systems; Analogy with Chemical Kinetics, Artif Life, 1:413-427 ©1995 MIT Press)

In the course of this self-assembly, various intermediate products are generated: monomers (isolated single units), dimers (two joined units), and so on up to hexamers, the final products. The following is a complete list of all conceivable joining reactions:

$$2\,X \rightarrow X_2\,, \qquad X + X_2 \rightarrow X_3\,,$$
$$X + X_3 \rightarrow X_4\,, \qquad X + X_4 \rightarrow X_5\,,$$
$$X + X_5 \rightarrow X_6\,, \qquad 2\,X_2 \rightarrow X_4\,,$$
$$X_2 + X_3 \rightarrow X_5\,, \qquad X_2 + X_4 \rightarrow X_6\,,$$
$$2\,X_3 \rightarrow X_6\,.$$

When the density (the number of units) of each cluster is

$$\mathbf{x} = (x_1, \dots, x_2)$$

and its dynamics are given as follows

$$\mathbf{x}(t+1) = \mathbf{x}(t) + \mathbf{F}(\mathbf{x}(t))\,,$$

by integrating these equations, the system behavior can be traced. The function F can be estimated relatively easily from the geometric shape (the angle range of collisions causing connection) of the units. Fig. 3.14 shows the results of numerical computation using such estimates.

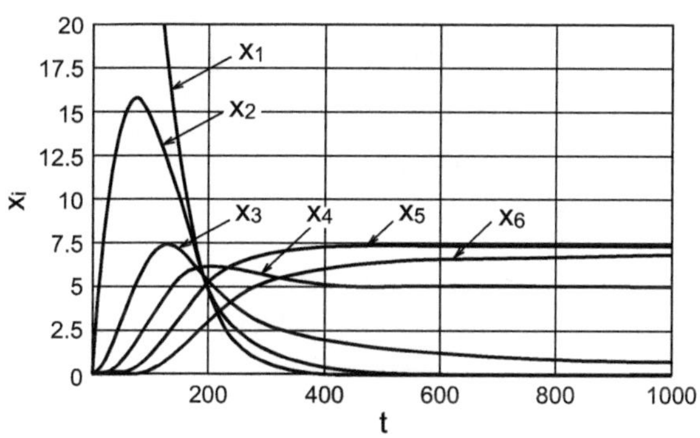

Fig. 3.14 Results of numerical analysis of self-assembly process (Hosokawa K, et al (1995) Dynamics of Self-Assembling Systems; Analogy with Chemical Kinetics, Artif Life, 1:413-427 ©1995 MIT Press)

Fig. 3.14 shows that, as soon as the agitation starts, many monomers become dimers, which then immediately join up with each other to become tetramers or join with monomers to become trimers, so that dimers are quickly consumed. But it also shows that there are substantial amounts of pentamers and tetramers that fail to become hexamers in the end. This is consistent with the results of actual agitation experiments.

How can the generation of undesired intermediate products be suppressed? Hosokawa's solution was to introduce the activation mechanism suggested by Penrose. One way to realize this mechanism is to allow the magnets to slide along a groove, so that only when magnets protrude from the surface (the activated state) can the units become connected, as shown in Fig.3.15. The self-assembly occurs only when a unit is activated by an external magnet, as in Fig. 3.15(d), and then becomes a seed. Simulations in which such a mechanism is introduced show that as many hexamers as the number of initial seeds were generated, with only monomers remaining. This means that the maximum possible yield was achieved.

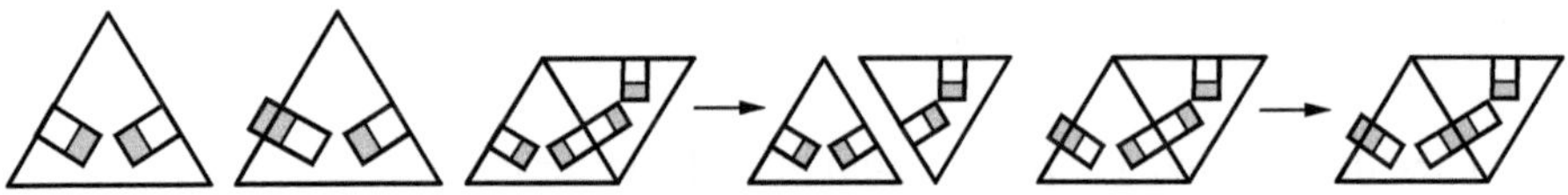

Fig. 3.15 Activation mechanism by sliding magnets (Hosokawa K, et al (1995) Dynamics of Self-Assembling Systems; Analogy with Chemical Kinetics, Artif Life, 1:413-427 ©1995 MIT Press)

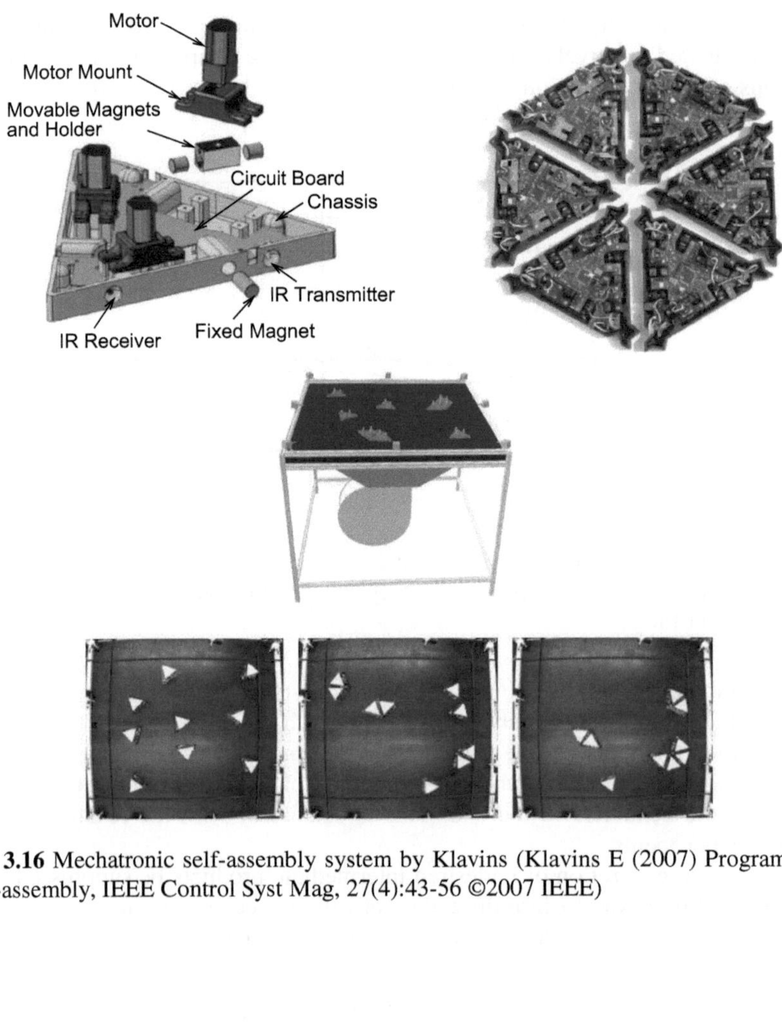

Fig. 3.16 Mechatronic self-assembly system by Klavins (Klavins E (2007) Programmable Self-assembly, IEEE Control Syst Mag, 27(4):43-56 ©2007 IEEE)

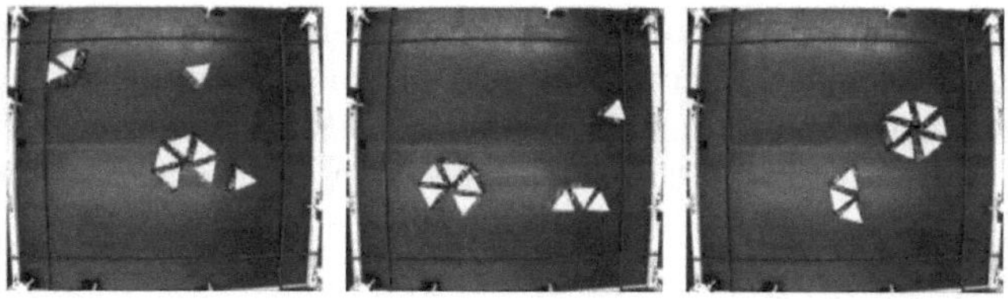

Fig. 3.16 *(continued)*

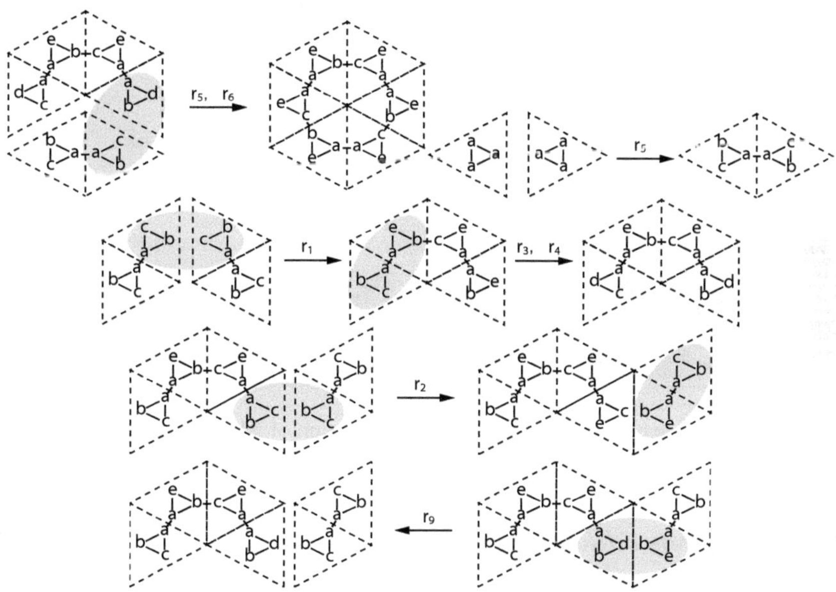

Fig. 3.17 Assembly rules to maximize the yield of hexagons (Klavins E (2007) Programmable Self-assembly, IEEE Control Syst Mag, 27(4):43-56 ©2007 IEEE)

3.4.2 Mechatronic Self-assembling System by Klavins

The model by Eric Klavins uses units which carry magnets like the ones by Hosokawa but which also are mounted with motors and microprocessors so that connections made by magnets can be disconnected [12]. Moreover, these units are capable of exchanging information with each other through digital communication. When these units are placed floating on an air hockey table and are randomly agitated by fans placed around the table, the units start colliding with each other. When the magnets of two units come close, those units become connected, and at the same time the two units communicate with each other. Based on the information obtained through the communication, they may cut the connection (Fig. 3.16).

Klavins considered when connections should be cut in order to generate more hexagons, and derived a set of rules (grammar) for maximizing the yield

(Fig. 3.17). To do this, he had to model the self-assembly process as a Markov process. The problem thus became how to increase the number of desired final products by directly controlling the relevant constants of the numerous reaction paths in the self assembly process. This research can serve as a foundation for constructing programmable chemical reaction systems.

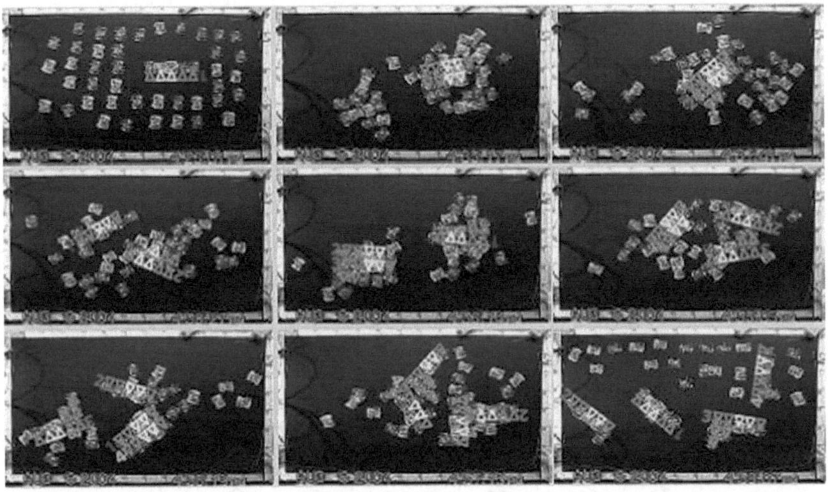

Fig. 3.18 Griffith's self-reproducing system (reproduction of 5 bit sequences) (Griffith S, et al (2005) Robotics: Self-replication from random parts, Nat 437:636 ©2005 NPG)

3.4.3 Self-reproducing System by Griffith

Saul Griffith constructed a self-reproducing system in which a chain of modules serving as a seed, together with many single modules are placed on an air hockey table and agitated, like the system by Klavins in the previous section. The system can produce a module chain that has exactly the same structure and the same information as the seed [13]. The modules are rectangular pieces and have ratchets on each side, which become connected upon collision. When such a connection is made, the microprocessors on the modules communicate with each other, investigating the internal state of each other's module. They are programmed to separate if the connection is found to be unnecessary.

According to Griffith, in order for a system to self-reproduce, there exist minimum requirements for the number of different types of modules and for the number of possible states of a module. In the case of biological systems consisting of proteins and nucleic acids, each molecule is a module having one state, and 24 different kinds of modules in total, 20 amino acids and four nucleic acids, are required. In the system of Griffith, self-assembly is possible with two different tiles in six different states [14].

References

[1] von Neumann, J., Burks, A. (eds.): Theory of Self-Reproducing Automata. Univ. Illinois Press (1966)

[2] Freitas Jr., R., Merkle, R.: Kinematic Self-Replicating Machines, Landes Bioscience (2004)

[3] Penrose, L.S.: Mechanics of Self-replication. Ann. Human Genetics 23, 59–72 (1958)

[4] Penrose, L.S.: Self-reproducing Machines. Sci. Am. 200, 105–114 (1959)

[5] McCulloch, W., Pitts, W.: A logical calculus of the ideas immanent in nervous activity. Bull. Math. Biophys. 7, 115–133 (1943)

[6] Turing, A.M.: The Chemical Basis of Morphogenesis. Philos. Trans. Royal Soc. Lond. B 237(641), 37–72 (1952)

[7] Schrödinger, E.: What is Life? Mind and Matter. Cambridge University Press (1944)

[8] Langton, C.G.: Self-reproduction in cellular automata. Physica D 10, 135–144 (1984)

[9] Tomita, K., et al.: Graph Automata; Natural Expression of Self-reproduction. Physica D 171(4), 197–210 (2002)

[10] Tomita, K., et al.: Automatic Generation of Self-Replicating Patterns in Graph Automata. Int. J. Bifur. Chaos. 16(4), 1011–1018 (2006)

[11] Hosokawa, K., et al.: Dynamics of Self-Assembling Systems; Analogy with Chemical Kinetics. Artif. Life 1, 413–427 (1995)

[12] Klavins, E.: Programmable Self-assembly. IEEE Control Syst. Mag. 27(4), 43–56 (2007)

[13] Griffith, S., et al.: Robotics: Self-replication from random parts. Nat. 437, 636 (2005)

[14] Griffith, S.: Growing Machines, PhD Thesis. MIT (2004)

Chapter 4
Basics in Mathematics and Distributed Algorithms

Abstract. In this chapter, we study briefly the mathematical foundations for building self-organizing mechanical systems. First, we discuss diffusion dynamics in a physical continuum, and then introduce reaction-diffusion systems that generate spatial patterns, and cellular automaton models which behave like those continuous systems in discrete state, space and time. Since the actual hardware realization of self-organizing mechanical systems often utilizes digital control and communication, some typical problems with such distributed information systems will be discussed.

4.1 Distributed System and Components

The behavior of an entire self-organizing system is determined by the characteristics of its individual components, their interactions, and external influences. For 'mechanical' systems, these factors are either physical (such as dynamic characteristics, interactions of forces, and disturbances), or informational (such as algorithms, communications, and sensor inputs). Most of the systems we discussed in Chapter 3 consist of components with simple functions that make mechanical connections, while those we introduce in the rest of the book use components equipped with microcomputers.

When microprocessors are used in components, the characteristics and interactions of components can be of a complexity limited only by the hardware of the processors, and it is also possible that each component functions differently from others. Having said that, in order to achieve the flexibility, the scalability, and the fault tolerance discussed in Chapter 1, the self-organizing systems are required to have components that are *homogeneous* (or slightly heterogeneous)[1], and their informational interactions should be local and symmetric. Our aim is to let components develop different functionalities through local interaction starting with asymmetrical initial conditions and boundary conditions (inputs from outside).

---

[1] Though the term 'homogeneous' refers to a whole system which is composed of and filled with the same components, it is often used in this book and other works in the phrase "homogeneous component", referring to a component of a homogeneous system. Similarly, the term 'slightly heterogeneous' here implies that the system is composed of a few types of components.

S. Murata and H. Kurokawa: Self-Organizing Robots, STAR 77, pp. 59–75.
springerlink.com

Since self-organizing mechanical systems are physical systems, any mechanical connections are fundamentally local. What the *locality* requirement actually means is that the interactions, or the informational connections, between components must be realized through the mechanical connections. On the other hand, information systems usually allow direct communications between components that are physically distant, and various network configurations such as those shown in Fig. 4.1 are used. However, in the case of the configuration shown in 4.1(a), addition of one new node requires introduction of connections with all existing nodes, and thus scalability is not realistically attainable. Furthermore, in the case of the configurations in Fig. 4.1(b), 4.1(c), 4.1(d), removal of one particular component breaks connectedness of the whole, so that these are not very fault tolerant. The bus connection which is frequently used in computer networks works like a fully connected topology as in Fig. 4.1(a) when the number of nodes is small. Since the nodes are only connected to the bus signal line, there is excellent scalability. As the number of nodes increases, however, this desirable property is not maintained and the system has to be structured into a hierarchical system of subsystems, each with hubs. Even in such case, scalability and fault tolerance are not easily maintained. Therefore, the locality requirement here is to limit the possible connection topology of graphs to be planar (Fig. 4.1(e)) or three dimensional, so that neighborhood relations in a graph are similar to neighborhoods in physical space.

The requirement of *symmetry* means that the components in any location have the same functions (homogeneity), and for each component the method of selecting connections is the same, as in Fig. 4.1(e), so that components interact in the

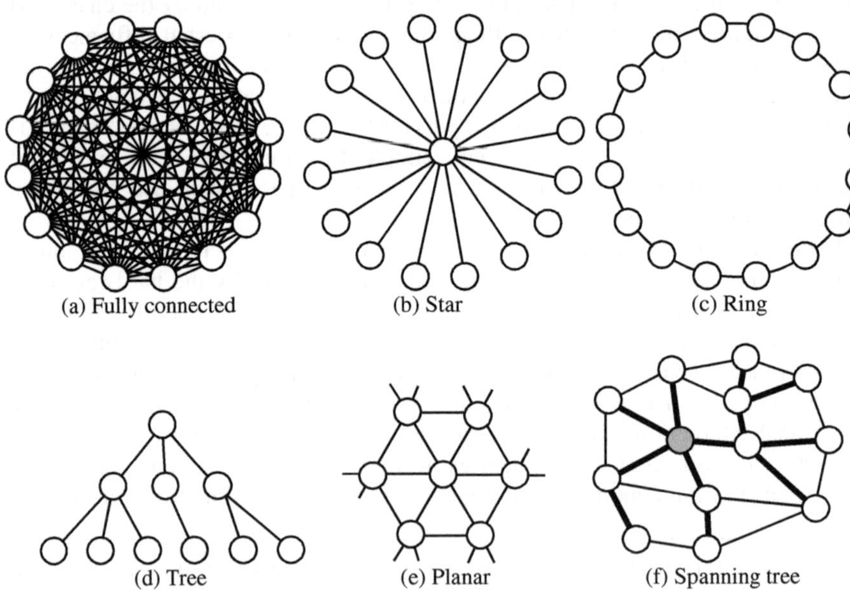

Fig. 4.1 Network topologies

same way in any direction. As the behavior of a deterministic system is uniquely determined if the components' functions and initial states are made homogeneous, we assume in self-organizing systems arbitrariness and variability of initial and boundary conditions or probabilistic behavior in each component's function.

When all the interactions between components are assumed to be local, there is no way for two distant components to be set to precisely the same time (*asynchrony*). Moreover, there is no way for them to immediately share information about the entire system (*inaccessibility to global information*). The symmetry requirement also dictates that components do not carry any information by which one can be identified from among the others (*anonymity*).

We aim for a system that organizes by itself through such symmetric local interaction. In fact, requirements such as locality, symmetry, asynchrony and anonymity are too strong for the entire system to develop specific desired functions. Therefore for example, in Chapter 8, when we discuss in detail the motion control of robots, some of these requirements are intentionally neglected or relaxed. In other cases, we try to meet these requirements as fully as possible and then find out what kind of global functions can emerge.

Hereafter, in order to gain some understanding of the behavior of systems that interact symmetrically and locally, we focus on a mathematical model of the diffusion process, one of the simplest processes in a continuous space. It can be regarded as the limit of a distributed system as the number of components increases. Diffusion is a process making the entire system uniform, but we will see that it can also generate some uneven patterns over the entire system under specific boundary conditions, or with dynamic characteristics of components.

4.2 Diffusion

4.2.1 Diffusion Equations

When a drop of ink falls in water, the color of the ink will gradually spread throughout the water. This is an example of the diffusion process, the most basic phenomena that are caused by local and symmetric interactions satisfying the law of conservation. A diffusion system is used as a model not only for physical phenomena such as heat transfer and chemical diffusion but also for consensus formation processes among numerous autonomous components. In the simplest case, a system of two components is described by the following mutual interactions.

$$\mathrm{d}s_1/\mathrm{d}t = D(s_2 - s_1),$$

$$\mathrm{d}s_2/\mathrm{d}t = D(s_1 - s_2)\,, \tag{4.1}$$

where s_1 and s_2 are variables (real number) representing component states, t represents the time, and the coefficient D is a positive constant. Since the difference $\Delta = s_1 - s_2$ of the two state variables satisfies

$$d\Delta/dt = -2D\Delta, \quad (4.2)$$

the difference Δ asymptotically approaches zero, which means that s_1 and s_2 converge to the same value as time passes. The convergence is faster when D is larger.

If we increase the number of components, place them in a straight line, and set the interactions between neighboring components to be the same, the following relation holds:

$$ds_i / dt = D((s_{i+1} - s_i) + (s_{i-1} - s_i)), \quad (4.3)$$

where the integer indices i indicate the order of the components in the line.

If we further increase the number of components to infinity, using the density $\rho(x)$ at a point x as the state variable, we obtain the following:

$$d\rho/dt = D\, d^2\rho/dx^2. \quad (4.4)$$

Extending this equation to two or three dimensions, we obtain the following diffusion equation:

$$d\rho/dt = D\, \nabla^2\rho, \quad (4.5)$$

where ∇^2 is the Laplacian. For spatial coordinates (x, y, z),

$$\nabla^2 = \partial^2/\partial x^2 + \partial^2/\partial y^2 + \partial^2/\partial z^2. \quad (4.6)$$

In any of the above cases, diffusion causes all the state variables to eventually converge to the global average value even while each approaches its local average value. By taking the sum or the integral of the entire system, the right hand side of the equation always becomes zero. Hence, the sum of s or the integral of ρ over the space is preserved (the law of conservation). The larger the diffusion coefficient D is, the farther and the more rapidly homogenization proceeds.

The systems described above only converge to uniformity. When there are sources and sinks in the system, however, there may be gradients of variables. Letting σ denote the volume of the flow at a source / sink per unit time (negative values if there is a sink):

$$d\rho/dt = D\, \nabla^2\rho + \sigma. \quad (4.7)$$

If the total sum of the flow is zero, there is an equilibrium state in which the left side of the equation is zero, namely

$$D\, \nabla^2\rho = -\sigma. \quad (4.8)$$

For example, in a system with a source and a sink with equal amounts of flow, any points other than these two points are in an equilibrium state described as

$$D\, \nabla^2\rho = 0, \quad (4.9)$$

and in the one dimensional case, the distribution of ρ becomes linear between the source and the sink.

In actual processes, as discussed in Section 2.5, it is often the case that a source and a sink with states fixed to constant values define the boundary condition of a closed system, yielding an equilibrium state with a linear gradient between the source and the sink.

4.2.2 Gradient Field

As we have seen, a diffusion system with a source and a sink creates a scalar field with non-zero gradient. The vector field of this gradient creates a flow field in the system. Using the nabla operator

$$\nabla = (\partial/\partial x, \partial/\partial y, \partial/\partial z) , \tag{4.10}$$

the gradient of a scalar field p is given as a vector field

$$\mathrm{grad}\, p = \nabla p = (\partial p/\partial x, \partial p/\partial y, \partial p/\partial z) . \tag{4.11}$$

The divergence of a vector field $\boldsymbol{u}$, given as

$$\mathrm{div}\, \mathbf{u} = \nabla \cdot \mathbf{u} = \partial u_x/\partial x + \partial u_y/\partial y + \partial u_z/\partial z , \tag{4.12}$$

is the sum of inflow into and outflow from a unit volume around a point (x, y, z). When p is a diffusion field, the divergence of its gradient is given as

$$\mathrm{div}\, (\mathrm{grad}\, p) = \nabla \cdot \nabla p = \nabla^2 p, \tag{4.13}$$

and from Eq. (4.8), it is equal to the amount of flow coming out of the source or going in the sink. Where there is no source or sink, the divergence of the vector field is zero, meaning that it represents incompressible flow.

In addition, the rotation, defined as

$$\mathrm{rot}\, \mathbf{u} = \nabla \times \mathbf{u} = (\partial u_z/\partial y - \partial u_y/\partial z, \partial u_x/\partial z - \partial u_z/\partial x, \partial u_y/\partial x - \partial u_x/\partial y) \tag{4.14}$$

satisfies the relation

$$\nabla \times (\mathrm{grad}\, p) = \nabla \times (\nabla p) = (\nabla \times \nabla) p = 0 , \tag{4.15}$$

showing that the gradient of any chosen scalar field is a vector field without a vortex, i.e., rotation of flow.

4.2.3 Pattern Formation by Reaction-Diffusion System

When multiple autonomous components which have their own dynamics interact by diffusion, the whole system becomes a reaction-diffusion system [1, 2]:

$$\mathrm{d}\rho/\mathrm{d}t = f(\rho, t) + D\, \nabla^2 \rho , \tag{4.16}$$

where the first term in the right is the autonomous (reaction) term, and the second is the diffusion term. This type of system is used as a model for chemistry and biology. Because of the reaction term, the entire system may show various behaviors. When the system without the diffusion term, the reaction system, has more than one stable equilibrium state, the system with diffusion may exhibit various patterns depending on the initial state. If the reaction system has a unique stable equilibrium point, however, the entire system settles at this equilibrium regardless of the initial conditions and characteristics of the reaction term. It seems clear that as diffusion puts the system into uniform equilibrium, combining stable reaction with diffusion results in a stable and uniform entire system.

However, things can be quite different when two state variables are involved. Alan Turing, who created the foundations of computer science, demonstrated that introduction of diffusion into a stable reaction system with two variables may lead to instability [3]. To illustrate this, we consider the following reaction system with two state variables u and v:

$$\mathrm{d}u/\mathrm{d}t = a\,u - b\,v,$$

$$\mathrm{d}v/\mathrm{d}t = c\,u - d\,v\,, \tag{4.17}$$

where a, b, c and d are positive coefficients.

Let us investigate the stability around the equilibrium point (0, 0) of this system. In general, the solutions to a system of linear differential equations like this are the combination of two solutions of the following form expressed with the complex eigenvalue λ and the exponential function exp(x):

$$(u, v) = \mathrm{Re}\,(\exp(\lambda\,(t+t_0))\,)(u_0, v_0)\,, \tag{4.18}$$

where Re(x) denotes the real part of a complex number x.

Letting $\lambda = p + q\,\mathrm{i}$ with i denoting the imaginary unit, we have

$$\mathrm{Re}(\exp(\lambda t)) = \exp(pt)\cos(qt),$$

so that the solutions will diverge when the real part p is positive and will converge to zero when p is negative, independently of the imaginary part q of λ. In other words, the equilibrium point is stable when the real parts of all the eigenvalues are negative.

In order to determine the eigenvalues, we substitute equation (4.18) for u and v in equation (4.17). Since differentiating the exponential with respect to t is equal to multiplying the exponential by λ, we obtain the following:

$$\lambda\,u_0 = a\,u_0 - b\,v_0,$$

$$\lambda\,v_0 = c\,u_0 - d\,v_0\,. \tag{4.19}$$

Here, in order for the non-zero solutions for (u_0, v_0) to exist, the determinant of the coefficient matrix of the following system of linear equations

$$0 = (a - \lambda)\,u_0 - b\,v_0$$

$$0 = c\,u_0 - (d + \lambda)\,v_0 \tag{4.20}$$

should be zero. Equivalently,

$$-(a-\lambda)(d+\lambda)+b\,c=\lambda^2-(a-d)\lambda+b\,c-a\,d=0 \tag{4.21}$$

should hold. The solution of this quadratic equation gives the eigenvalues λ. The relation between the roots and the coefficients here dictates that the sum of the two eigenvalues is $(a-d)$ and their product is $bc-ad$. This shows that whether the eigenvalues are real or complex numbers, the conditions for both of their real parts to be negative are

$$0<a<d, \quad a\,d<b\,c\,. \tag{4.22}$$

When these conditions are satisfied, if the eigenvalues are real numbers the convergence is asymptotic, while if they are complex numbers the convergence is oscillatory (Fig. 4.2(a) and (b)).

Now, we consider two systems of (u_1, v_1) and (u_2, v_2) that satisfy equations (4.17) and (4.22), and join these two stable systems via the following diffusion interactions of u and v:

$$\mathrm{d}u_i/\mathrm{d}t=a\,u_i-b\,v_i+D_u(u_j-u_i),$$

$$\mathrm{d}v_i/\mathrm{d}t=c\,u_i-d\,v_i+D_v(v_j-v_i), \tag{4.23}$$

where the indices (i, j) are either (1, 2) or (2, 1). If we add up equations (4.23) with regard to u and v respectively, the diffusion terms are cancelled out. Equivalently, by substituting

$$U=u_1+u_2, \quad V=v_1+v_2\,, \tag{4.24}$$

one can obtain equation (4.17), which indicates that (U, V) is stable around (0, 0).

On the other hand, regarding the new variables

$$\Delta u=u_1-u_2, \quad \Delta v=v_1-v_2\,, \tag{4.25}$$

the following equations hold:

$$\mathrm{d}\Delta u/\mathrm{d}t=(a-2D_u)\Delta u-b\,\Delta v\,,$$

$$\mathrm{d}\Delta v/\mathrm{d}t=c\,\Delta u-(d+2D_v)\Delta v\,. \tag{4.26}$$

The first stability condition in (4.22) of this new reaction-diffusion system is satisfied if $a>2D_u$ holds. However, when D_v is sufficiently large, we have

$$(a-2D_u)(d+2D_v)>b\,c\,, \tag{4.27}$$

and therefore the second condition in (4.22) is not satisfied, leading to instability. In this case, a little fluctuation from the equilibrium state will cause Δu and Δv, or equivalently, u_i and v_i, to diverge as time passes. This phenomenon is called *Turing instability*.

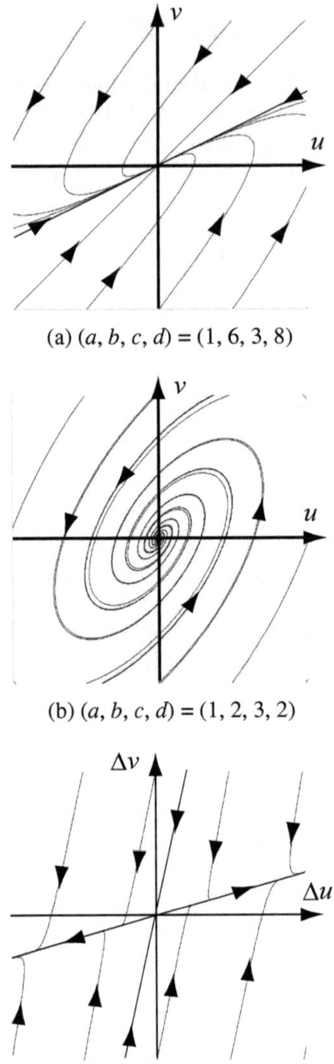

(a) $(a, b, c, d) = (1, 6, 3, 8)$

(b) $(a, b, c, d) = (1, 2, 3, 2)$

(c) Instability by diffusion

Fig. 4.2 Behavior near the stable equilibrium point. (c) is the behavior of the system (4.23) with the reaction terms of (b) and diffusion terms D_u=0.1, D_v=4.

Note that in case of linear reaction terms as in (4.23) the divergence will progress to infinity, but in general physical or chemical systems there is non-linearity in the system that prevents divergence. For instance, if we replace au in the reaction term with $au(1 - hu^2)$, the system will behave just as the case of au in the neighborhood of (0, 0), but it does not diverge because $-ahu^3$ grows rapidly as u becomes larger.

Let's look at this instability in a continuous space. Consider the following one dimensional reaction-diffusion equations of density variables u and v on axis x with the same reaction terms as above:

$$du/dt = au - bv + D_u d^2u/dx^2 ,$$

$$dv/dt = cu - dv + D_v d^2v/dx^2 . \tag{4.28}$$

The conditions for stabilization without the diffusion terms are also given by (4.22) for this case. Now, we add to the equilibrium state a small fluctuation:

$$(u, v) = \mathrm{Re}\,\exp(i\,k(x+x_0))\,(U, V) , \tag{4.29}$$

where U and V are functions of time t only. Equation (4.29) indicates that the graphs of u, v are sine curves with amplitude U or V and wavenumber k (wavelength $2\pi/k$) (Fig.4.3). Substituting this for u and v in (4.28) just as before, we obtain the following:

$$dU/dt = (a - D_u\,k^2)U - bV ,$$

$$dV/dt = c\,U - (d + D_v\,k^2)\,V . \tag{4.30}$$

When D_u is sufficiently small and $(a - D_u\,k^2) > 0$ holds, the first condition in (4.22) is satisfied. The second stability condition is expressed as

$$(a - D_u\,k^2)(d + D_v\,k^2) < b\,c . \tag{4.31}$$

When D_v is sufficiently large, the left hand side becomes larger than bc near the wavenumber k_c as seen in Fig.4.4, and therefore the inequality does not hold. Thus, fluctuations in such wavenumber grow like a wave pattern. In the case of two dimensional space, it appears as a stripe pattern called a *Turing pattern*.

In equation (4.17), u is called *activator* because when u increases, both u and v increase, while v is called *inhibitor* because increase in v will result in decrease of u and v. Diffusion coefficients satisfying $D_u \ll 1$ and $D_v \gg 1$ will lead to instability. In short, the Turing pattern is generated when the activators work locally and the inhibitors work at a distance.

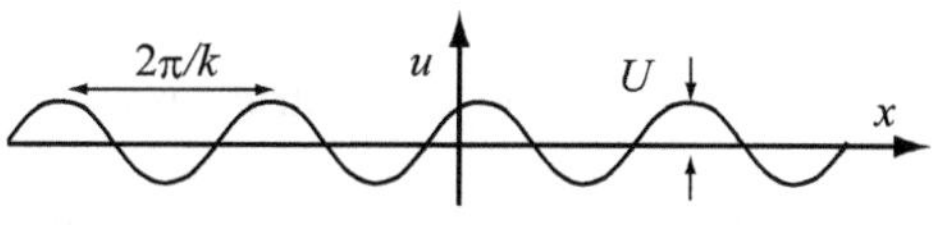

Fig. 4.3 Sinusoidal wave fluctuation

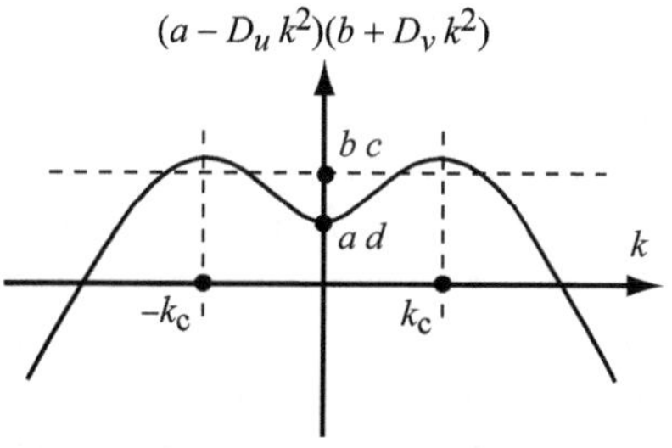

Fig. 4.4 Wave length and stability measure

Let us explain the mechanism of pattern generation qualitatively again. Assume that the system is in an equilibrium state, and that at some point an activator u increases locally by a small amount (Fig. 4.5(a)). This activation results in increase of both u and v. In the case without diffusion, the increased v in turn works to suppress the further increase of u and v, and the system stabilizes at the equilibrium again. But with diffusion, since the inhibitor diffuses away faster than the activator, inhibition at the site of the activator is insufficient to stop the increase of the activator. Meanwhile, in the neighborhood of this site, both the inhibitor and the activator spread through diffusion. Since the inhibitor diffuses before the activator does, both are inhibited and the activator does not increase. As a result, an undulating shape as shown in Fig.4.5 appears. If there are plural peaks, their surrounding plains interfere with each other, and peaks do not approach each other closer than a certain minimum distance. On the other hand, where peaks are far apart, new peaks will appear when there are fluctuations. Thus, a state of equilibrium will be reached in which there is a spatial wave pattern (Fig. 4.5(c)).

Fig. 4.5 Generation process of pattern

The above described a case of a stationery pattern in an equilibrium state. Below is a system that presents a similar pattern that travels across the space:

$$du/dt = u(1-u)(u-a) - v\,,$$

$$dv/dt = \varepsilon(u - \gamma v)\,, \tag{4.32}$$

where $0 < a < 1/2$, $\gamma > 0$, and $0 < \varepsilon << 1$. Although (0, 0) is a stable equilibrium, this system does not behave as simply as the case of Fig. 4.2 due to its nonlinearity. Instead, depending on its initial state, it may exhibit impulsive behavior, growing for a while and then settling back to the origin. If we add diffusion to this system, a stimulus given to a single point induces an excited state which travels like waves. This system is different from the Turing pattern which is generated by instability. When an initial spiral pattern is introduced, the system sustains a spiral pattern which expands outward as it rotates (it is possible to generate patterns like the one in Fig. 4.8).

A particularly famous example of various behaviors as above is the Belousov-Zhabotinsky chemical reaction (BZ reaction). This system exhibits phenomena such as oscillating color changes of a solution in a beaker or concentric circles and spiral color patterns in a Petri dish solution.

The above models for the generation of static or dynamic space patterns may account for the generation of various shapes and skin patterns of biological organisms (See Section 2.5).

4.3 Cellular Automata

Discretization of space and time of the above systems is done to transform a continuous system described by differential equations to one described by difference equations. If we go further and set each of these discrete segments to have a limited number of discrete states (finite state), we obtain cellular automata [4, 5].

A cellular automaton consists of cells aligned in a grid, which change their states according to their interactions with neighboring cells. The notions of neighborhood typically used in the case of two dimensional square lattices are the *von Neumann neighborhood*, which consists of the four cells neighboring the central cell on each of the four sides, and the *Moore neighborhood*, consisting of the eight cells surrounding the central cell. The states of all cells are updated synchronously at discrete time intervals. State transition rules determine the state of a cell in the next time based on the current state of the cell and those of neighboring cells. It is possible to use cellular automata to model a physical, chemical, or biological system, but it is also possible to create a system that behaves completely different from any physically real phenomena.

4.3.1 Field of Diffusion

If we change our one dimensional diffusion equation (4.3) to have discrete spatial positions and discrete times, the following difference equation is obtained:

$$s_i(t+1) = s_i(t) + D\left((s_{i+1}(t) - s_i(t)) + (s_{i-1}(t) - s_i(t))\right) + \sigma . \qquad (4.33)$$

Since a cellular automaton has only a finite number of states, s is kept inside this range.

For an arbitrary dimension, given a cell i and its neighborhood j, let the difference Δ be

$$\Delta_i = \Sigma_j (s_j(t) - s_i(t)) . \qquad (4.34)$$

Then, by simplifying equation (4.33) as below, we obtain a pseudo-diffusion system:

$$s_i(t+1) = \begin{cases} s_i(t)+1, & \Delta_i > 1, \\ s_i(t), & \Delta_i = 1, 0, -1, \\ s_i(t)+1, & \Delta_i < -1, \end{cases} \qquad (4.35)$$

In this system, sources and sinks can be created by fixing the values of some state variables to the maximum or the minimum.

Another similar pseudo-diffusion system is given by

$$s_i(t+1) = \min_j s_j(t) + 1 , \qquad (4.36)$$

where the first term on the right hand side is the minimum value among the states of a cell and its neighborhood cells. By this rule, the state of each cell becomes greater at most by 1 than the minimum of the states of cells in its neighborhood, and therefore, if there is a sink fixed at $s = 0$, then the state of a cell at distance d from the sink is no more than d. In other words, if a cell has a state s, it means that there is no sink among the cells located within the circle around that cell with radius s. Conversely, by using a sink, the states of cells located within radius r from the sink are kept less than r. This mechanism is useful for suppressing the behavior of surrounding components in the mechanical systems discussed in the chapters that follow (see Section 5.5.2).

Since diffusion systems with discrete states as described above have the quantization problem discussed in the following section, in order for the diffusion effect to reach the entire system, it is in general necessary to increase the number of states as the size of the cell space increases.

4.3.2 *Flow Field*

We showed in the previous section that the gradient (grad operation) of the equilibrium states of a diffusion field can represent incompressible flow. In Chapter 8, we need to introduce a flow field in a cellular system. Is it always possible to generate a flow field using a diffusion type of cellular automaton?

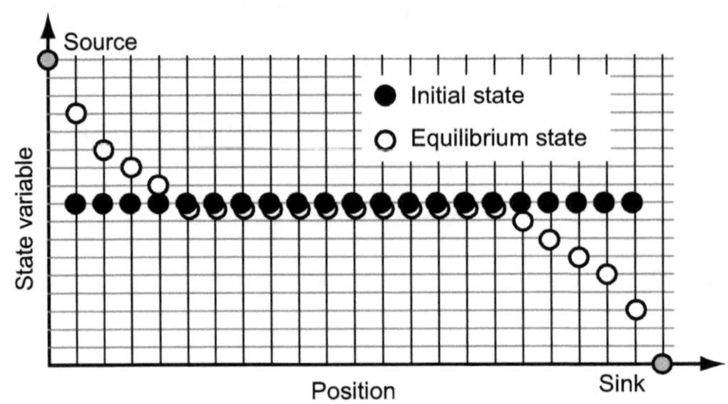

Fig. 4.6 A problem with discrete diffusion field. In equilibrium state, gradient is zero except in some neighborhoods of the source and the sink.

For example, consider a one dimensional finite cell space with uniform initial states. We assume a source and a sink placed at the ends, as shown by grey circles in Fig. 4.6. In this case, the effect of diffusion is strong near the source and the sink, and is weaker in points away from them. In a system with continuous states, however small the effect is, it is sufficient to cause some change in the states as time passes, and the system eventually reaches an equilibrium state with a uniform gradient. However, in a system with discrete states, at locations distant from the source and the sink, state transitions do not happen if difference from the neighboring states is insufficient. This means that except for areas near the source and the sink, it is impossible to generate a vector field of flow.

An effective way to resolve this problem is to magnify the state variables periodically and to introduce aperiodicity or probabilistic characteristics. In the simulation shown in Fig. 4.7, the difference between state variables of neighbors is calculated in modulo, and updating is carried out probabilistically. As a result, a gradient in the modulo is generated with initial and boundary conditions similar to Fig. 4.6. It is also possible to design a cellular automaton which generates a vector field directly. Note that such discrete systems may not have those properties such as zero-divergence or zero-vortex that a continuous system would have, and that in some cases they may oscillate or a vortex may remain.

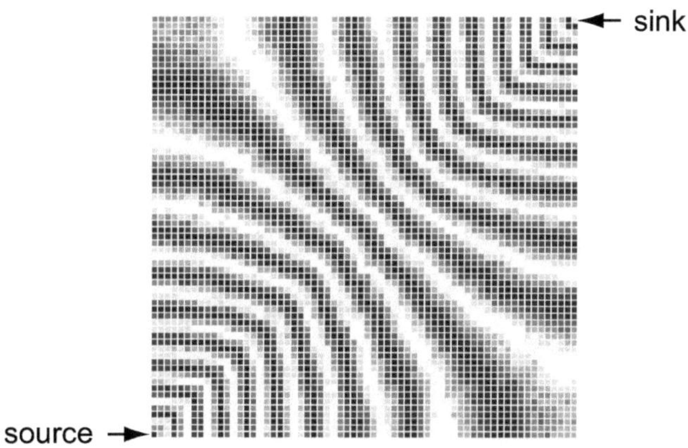

Fig. 4.7 Simulation of diffusion realized by expanding state variables. State variables are integers ranging from 0 (white) to 15 (black).

4.3.3 Game of Life

A particularly well known cellular automaton is the *Game of life*. In the Game of life, each cell has two possible states, which are usually represented by white and black cells and are called *dead* or *alive*. The state transition depends on the number of live cells in the Moore neighborhood, and follows these rules:

- **Reproduction:** any dead cell neighbored by exactly three live cells becomes a live cell
- **Sustainment:** any live cell neighbored by two or three live cells stays alive
- **Death:** any cell other than the above two cases dies.

Although this system has only two possible states and is driven by simple rules, it can generate a wide variety of patterns including ones that oscillate or resemble biological reproduction, depending on the initial pattern. For further details, please read the references or do experiments with computer programs by yourself.

It is also possible to create a system that generates dynamic patterns like BZ reactions by extending the Game of Life [6, 7]. For example, we can extend the possible states to integers from 0 to n; 0 is alive, from 1 to $n-1$ represents infected, and n is sick. The state transition rules are as follows:

- Any live cell (state 0) changes its state to $[A/k_1] + [B/k_2]$
- Any infected cell (state 1, ..., $n-1$) changes its state to $[(s_i + \Sigma s_j)/A] + g$
- Any sick cell (state n) becomes a live cell (state 0)

where A and B are the number of infected and sick cells in the neighborhood, respectively, k_1, k_2, and g are positive parameters, and s_i, s_j are states of the cell and its neighbors, respectively. The $[x]$ carries out the operation within the brackets upon the real numbers, and rounds the result down to an integer. If the obtained value is greater than n, n is made the result, but otherwise the obtained value is made the result. Depending on the value of g, the system generates dynamic patterns that oscillate, rotate as swirls moving outward, etc. (Fig. 4.8).

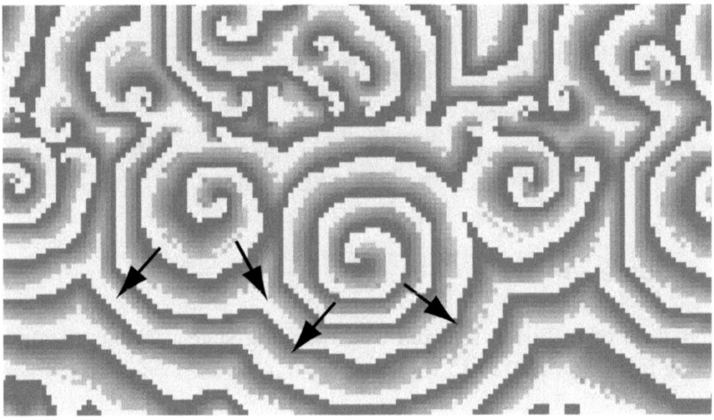

Fig. 4.8 A simulation example

4.4 Distributed Algorithms

In the case of physical phenomena such as a diffusion process, or the case of standard cellular automata, it is assumed that the entire system is synchronized with unique time and a local state variable is passed to all cells in the neighborhood at the same time, without fail. However, when components are physically implemented, things do not go as assumed in the ideal case, and various problems that are common in distributed information systems arise due to asynchrony, and delays and errors in communications. Distributed algorithms are introduced for resolving such problems of distributed information systems. In this section we explain some typical examples of such problems. Here we follow the standard terminology for information systems and refer to a component as a *process*, and to interactions between components as *communications* [8, 9].

4.4.1 Leader Election

Even with a distributed system, it is sometimes necessary to select a special process for centralized problem solving. In the following chapters, we will encounter the cases where system control is to be centralized. In other words, from among similar processes one is chosen which controls all the others. This is called *leader election*. When each process carries a unique identifier (ID), this problem can be reduced to that of finding the minimum or maximum value, for which various algorithms have been proposed.

For instance, in order to find the maximum ID, first each process sets its ID number as the value of a variable x. Then, two processes compare their x and select the one with the larger x as the winner. The loser set its x to that of the winner. By carrying out this comparison procedure for all neighboring pairs initially and every time there is an update, eventually all the processes will have the same x. Then, the process whose ID is the same as x is chosen as the leader. Even in the case where no ID is available, if the processes can generate random numbers which are different from each other, it is possible to select a leader using the same algorithm.

However, if the total number of processes is unknown, there is no way to tell when the selection process is complete. In many mechanical systems there is no need to choose a single leader for the entire system, as long as the communication neighborhood coincides with the range of physical interactions. In that case it suffices to choose a leader from among processes within a certain distance and carry out centralized processing within that range.

Although selecting a leader is beneficial to ensure system-wide consistency, an advantage of a centralized system, the system may lose advantages of a distributed system such as efficiency, fault tolerance and scalability. Therefore, the system should resume distributed control once the task that required centralized processing is over.

4.4.2 Spanning Tree Construction Problem

In a distributed system, message relaying is necessary for communication between two processes that do not have a direct link. If the route from the sender process to the receiver process is unknown, the only thing the sender can do is to broadcast the message and wait for the response. If processes receiving this message then broadcast it, this will result in cumulative proliferation of this same message in the system. In order to avoid this, it is necessary to introduce a mechanism to restrict duplicated messages or to limit the life time or travel distance of a message.

If a particular route for communication is fixed in advance, communication between remote processes becomes much easier. When the whole system is represented as a graph, in which a process and a communication path are represented by a node and a link respectively, the problem is to find a tree structure without a loop that connects all the nodes. Such a tree is called a *spanning tree*. As shown by the heavy lines in Fig. 4.1(f), in a spanning tree a path between any two nodes is uniquely determined. Once a spanning tree is constructed, it is possible to introduce a hierarchical control structure over the whole system. The spanning tree is also used for solving problems such as the *broadcast problem* of sending the same data to all processes, the initialization problem where one of the processes transforms all the others into a particular state, and the leader election problem discussed in the previous section.

As is obvious from Fig. 4.1(f), however, communication relying only on the spanning tree lessens the advantage of distributed systems in speed and reliability. Moreover, in the mechanical systems described in the following chapters, the physical connections between components change with time, and the structure of the connections among processes is not fixed either. In such cases, communication using the spanning tree has certain limitations.

4.4.3 Exclusion Control

Assume that a resource is shared by several processes, but it is available only for one process at a time. In such a case, it is necessary to grant one process an exclusive access to the resource, while forbidding access by other processes. Such a mechanism is called *exclusion control*, or *mutual exclusion control*. If exclusion control does not work, *deadlocks* may occur, as we explain below, or there may be *starvation* of a process that is never granted access to resources.

4.4.4 Deadlock

Assume that two resources A and B are used exclusively by a process, and that processes 1 and 2 both require both resources A and B. If process 1 takes resource A while process 2 takes resource B, then both the processes will have to wait forever for the other resource to be released. Such a situation is called *deadlock*, a phenomenon where multiple processes wait for events that can never happen. In this example, if either of the processes gives up and releases the resource, the deadlock will be resolved.

Even for a single processor system, deadlocks are common bugs found in a program that consists of several subprograms. Deadlocks happen often in distributed systems but it is usually difficult to detect them. There are various algorithms for detecting deadlocks.

4.4.5 Reliability

Reliability both in process behavior and in communication between processes is crucial for achieving good system performance and resolving the issues discussed above. Network buses, which are widely used in computer networks, employ various techniques to cope with communication failures caused by message collision. For hardware, a method called TMR (Triple Modular Redundancy) is often used to obtain reliable results by unreliable components; three components are used in parallel and the output is determined by majority voting. There is also a method called N-version programming in which different programs configure the same function, similar to TMR, to reduce the impact of errors (bugs) in a program. The dynamic redundancy method reduces incidence of faults by correlating measurements of sensors for different quantities, such as angle, angular velocity, and angular acceleration, based on their dynamic relations. These and many other methods for improving the reliability of computing systems and control systems (fault-tolerant design) have been proposed.

These methods often require conditions such as the reliability of each component being high enough for there to be small probability of multiple simultaneous failures in the system, or the failure of a sensor causing its output not to become arbitrary but rather to be fixed to a predefined value. For a software process, these conditions are difficult to fulfill, and the above methods are not always effective. The Byzantine Agreement Problem (or Byzantine Generals Problem) for distributed systems asks how to obtain a correct consensus for action by the entire system when some processes fail or malfunction. Several algorithms operating under various assumptions have been suggested. For further details, please refer to [8, 9].

References

[1] Grindrod, P.: Patterns and Waves: The Theory and Applications of Reaction-Diffusion Equations. Clarendon Press (1991)

[2] Murray, J.D.: Mathematical biology II: Spatial models and biomedical applications, 3rd edn. Springer (2003)

[3] Turing, A.M.: Phil. Trans. R. Soc. Lond. B 237, 37–72 (1952)

[4] von Neumann, J.: The general and logical theory of automata. In: Jeffress, L.A. (ed.) Cerebral Mechanisms in Behavior: The Hixon Symp., pp. 1–31. Wiley, New York (1951)

[5] Ilachinsky, A.: Cellular Automata. World Sci. (2001)

[6] Gerhardt, M., Schuster, H.: A cellular automaton describing the formation of spatially ordered structures in chemical systems. Physica D 36, 209–221 (1989)

[7] Dewdney, K.: The hodgepodge machine makes waves. Sci. Am. 104 (1988)

[8] Lynch, N.A.: Distributed Algorithms. Morgan Kaufmann (1996)

[9] Tel, G.: Introduction to Distributed Algorithms. Cambridge Uni. (1994)

Chapter 5
Artificial Self-assembly and Self-repair

Abstract. In Chapter 2, we learned that biological systems have elaborate designs that cannot be matched by conventional engineering, thanks to the various self-organizing mechanisms deployed in each layer of their hierarchy. In Chapter 3, we summarized several attempts to reconstruct such biological mechanisms artificially. If we can uncover the principles of these mechanisms through reconstruction efforts, then we can in turn apply these principles in making devices for our own purposes. In other words, we proceed from science to engineering. Even if we obtain ideas for basic notions and principles from biological systems, though, when we actually build devices, many practical issues remain to be resolved. In such specific problem solving stages, there is no need to look to nature for ideas anymore; although humans were inspired by birds to create flying machines, the actual airplanes constructed by humans have quite different structures from those of birds. The technology or materials currently at hand must be used, and the problems may be solved with new ideas. The aim of this chapter is to construct artificial systems like the ones we discussed in Chapter 3, but this time from an engineering perspective. Specifically, we consider how to make systems which have self-assembly and self-repair functions.

5.1 Methods for Self-assembly and Self-repair: Homogeneous System Approach

The unit machine called a "*Fractum*", which we discuss in this chapter, is one typical realization of a distributed autonomous system [1-3]. All aspects of its hardware and algorithms are strictly based on the principles of distributed autonomous systems, such as homogeneity and local communication. Using such units, we built a mechanical system that is capable of self-assembly and self-repair. As we have seen in Chapter 2, biological self-repair depends on the self-reproduction of cells, and the technology of current ordinary mechanical engineering is not likely to be able to realize a system whose components can self-reproduce. However, we don't think that an artificial self-repairing system cannot be achieved until technology for self-reproduction of elements is completed. This obstacle can be hurdled if the conceptual framework of the system is changed appropriately.

S. Murata and H. Kurokawa: Self-Organizing Robots, STAR 77, pp. 77–103.
springerlink.com

Firstly, *even if components cannot self-reproduce themselves, it is sufficient that only one kind of component is used*[1]. Having only one kind of component is an advantage for self-repair because any broken component can be replaced by any other.

Secondly, *the components should have the ability to assemble into a system by themselves*. Here, the task of assembly can be further decomposed into two functions. One is the *mobility* of the component, its ability to move to a suitable position. In ordinary machine assembly, a human or a robot would move components to appropriate positions to assemble them into a machine. In the case of self-assembly by components, however, they themselves must move to appropriate positions by some method[2].

Another function required is *connectivity*. If components are to arrange themselves into a certain configuration to form a mechanical system, they must remain at certain geometric positions relative to each other. For that purpose, at least three connections are required for each component. If each component can connect to only two other components, the systems can only have a one-dimensional linear structure. If on the other hand each component can have at least three connections, desired three-dimensional structures can be built just like chemists build models of molecules.

Now, if we assume that the components already have connections when the assembly begins, the relative positions of components can be changed by *reconfiguring* the connections ("reconfiguration" hereafter means "replacing connections"). This means that the mobility of components is equivalent to their reconfiguration function.

To make components with such functions, each component needs control logic, and it must also have a certain level of information processing capability. Since the information processing in this case is autonomous and distributed, communication between units is also necessary.

So far, we have been using the term *component*, but realizing the above functions with simple components such as screws and magnets is out of the question. What we mean here by components is the unit of self-organization we discussed in Section 1.2. Therefore, in this chapter, we call such components *units*. (In later chapters, we call more advanced constituent elements *modules*. There is no strict

---

1 This is not a particularly new idea. The cellular model of von Neumann is also based on the assumption of "a space filled with homogeneous cells that can become any type" It should be kept in mind that, in this model, a cell of a particular type can differentiate into 29 possible types. This is in effect the same as assuming that the desired component is available at the desired position any time.

2 The best way to move a component depends on the size of that component (See Section 1.3.2). In von Neumann's kinetic model and Penrose's building block model, components are assumed to be floating in space, and when they randomly collide it is decided whether to use this connection in the assembly. However, this kind of assembly by random collision can be adopted only when the components are very small and the effect of the gravity can be ignored, not for ordinary sized mechanical systems. On the other hand, if we equip units with legs or wheels, the system will be reduced to a collection of mobile robots.

distinction between unit and module, but in general a unit is considered to be a comparatively primitive component.)

To summarize, our approach features:

1. Homogeneous architecture: the entire mechanical system is made of one kind of mechanical unit. Namely, all the units have the same hardware and the same software.
2. Connection capability of units: each unit can connect to multiple (three or more) units.
3. Connection reconfigurability: each unit has an actuator and can act in concert with other units to change the connections between them.
4. Information processing and communication capability: each unit has a simple information processing mechanism and can communicate with other units.

If such units are used in building a mechanical system, it is possible not only that a faulty unit can be replaced by any other unit but also that the configuration and the function of the whole system adjusts to the environment. Moreover, the assembly of the system does not require any external help but can be done by the units autonomously.

Fig. 5.1 describes our notion of self-assembly and self-repair by a machine based on this approach [1]. Let us first explain the notion of self-assembly. By *self-assembly* we mean units moving themselves into a target configuration by changing their relative positions. For instance, consider a collection of units that looks like a lump of clay with indeterminate shape as described in this diagram. The process in which the lump transforms into the target configuration after it is given information on the target configuration is called self-assembly. Here, the collection can have any initial configuration; since a collection of units in one shape (a chair, for example) transforming into another shape (say, a table) also is self-assembly. The liquid metal robot in the film "Terminator 2" is a good example.

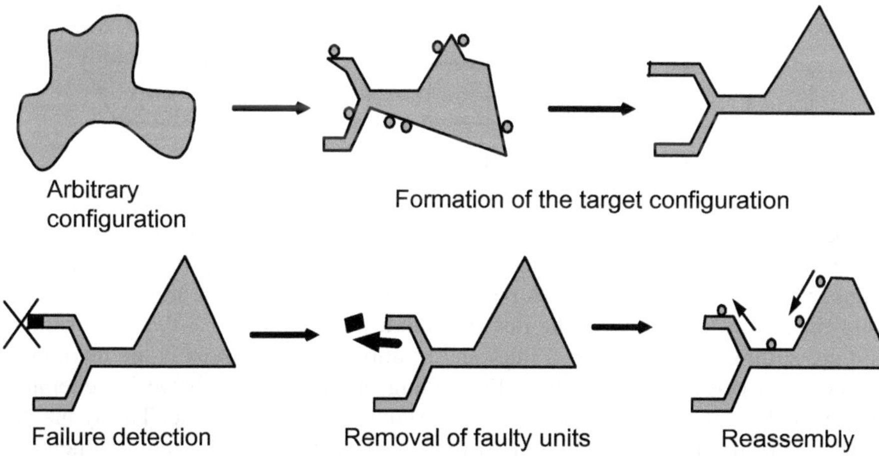

Fig. 5.1 Self-assembly and self-repair

On the other hand, *self-repair* is the process whereby after some units of a self-assembled system become defective, those faulty units are detected and removed, and the system is reconstructed by delivering spare units to replace the faulty ones. Spare units can be either stored in a designated part of the system or scattered throughout the system.

5.2 Hardware for Two Dimensional Units

From the discussion so far, we know that if we use the same units in constructing a mechanical system, self-assembly and self-repair are possible in principle. The next question is how to realize such units. Since our goal is to check the feasibility of the concept, we would like the hardware to be as simple as possible. Therefore, we consider two dimensional units. In two-dimensional space, the directions of connections and motions are restricted, which reduces the degree of freedom (the number of actuators). The algorithms for self-assembly and self-repair also become much simpler than the three-dimensional case.

The first thing to consider is the unit shape. Since we use the same units, packing them side by side in a two dimensional space (plane) would limit their movement. Crystalline lattices naturally come to mind, but given that we need to allow the units to have mobility, the unit shape should be geometrically symmetric (because if a unit has an asymmetric shape, the paths it can take without geometric interference would differ according to its spatial orientation). Tessellations made up of symmetric shapes, regular polygons in particular, can be made only in three ways as shown in Fig. 5.2. These are our starting points for our designs of the unit shape.

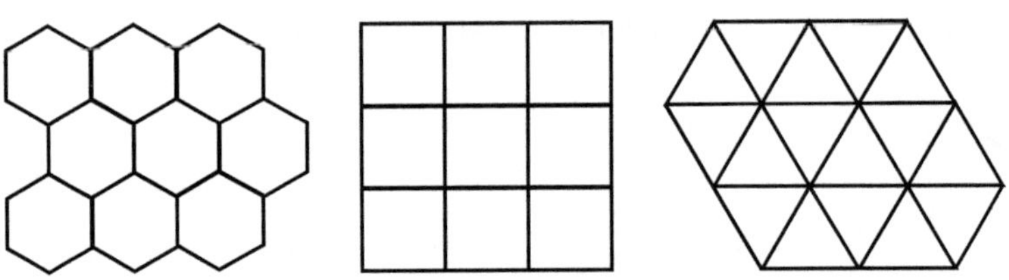

Fig. 5.2 Triangular, cubic and hexagonal lattices

Units are more reliable when they have a smaller number of parts and a simpler structure. The complexity of the unit structure is determined by its symmetry (the number of necessary connections per unit) and its mobility (the number of directions that a unit can move). The possible modes of motion for a two dimensional unit are sliding (movement in a line) and rotation. Another issue is the amount of torque generated by the actuators. Depending on how much torque is generated, only a single unit or multiple units are allowed to move at one time. The symmetry of units constrains the connection structures of two units to be complementary. We will discuss these issues in Chapter 6.

We chose a hexagonal lattice (Fig. 5.3) in order to satisfy the above requirements. In the following design, simple actuators realize both connection and locomotion by rotation. Also, units with this shape have a beautiful complementarity among themselves. We call this unit a *Fractum* (Fracta is the plural), standing for "fundamental component that constitutes a mechanical system".

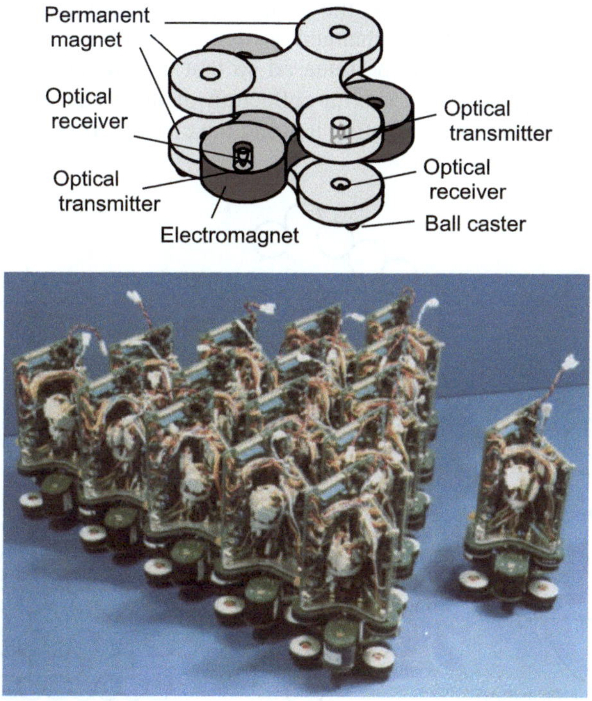

Fig. 5.3 Two dimensional unit Fractum

This unit is placed on a flat surface and is supported by ball casters but it cannot move by itself. The body has a three layer structure, each layer of which has a special shape consisting of combinations of arcs. There is no internal moving part. Each of the three layers has exactly the same shape, although the middle layer is rotated 60 degrees from the other two. Three permanent magnets are embedded in both the top and the bottom layers, the north pole facing upward. In the middle layer, instead of permanent magnets, electromagnets are embedded. Each unit has its own CPU, which controls the polarity of the electromagnets.

The basic mechanism for unit connection and locomotion is explained in Fig. 5.4. When the polarity of the electromagnets is set to be the same as that of the permanent magnets, an attracting force is generated. This force is used for connection. To release this connection, the polarity of the electromagnets is reversed. When an electromagnet is pulled in between permanent magnets, the magnetic field is constant even when the latter are rotating relative to the former. Thus, one unit can be used as a bearing for rotation of another unit. Moreover, just like self-aligning bearings, it has an automatic adjustment function because the center of each magnet works as the axis of rotation, and hence no additional positioning is required. The units are also capable of making connections and reconfiguring those connections. A unit can connect to as many as six other units, and by switching the electromagnets' polarities, the relative positions of two units which are connected to each other can be changed so that the units rotate relative to each other. Also, the connection can be released so that units detach from each other. By repeating these procedures, such a collection of units can transport a unit on the border of the collection by rotating it along the circumference.

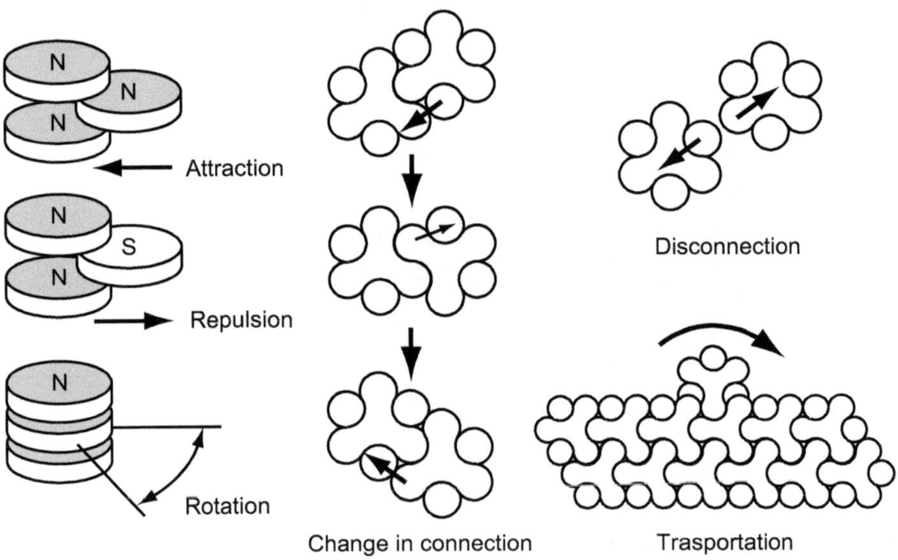

Fig. 5.4 Basic mechanism for connection and motion of units

Finally, we briefly explain the implementation of communication and information processing functions of the units. In this model, we use 8 bit microcomputers as information processing CPUs. Local information exchange between adjacent units is done with infrared communication devices which allow non-contact serial communication. An optical transmitter (an LED element) and an optical receiver (a photo transistor) are embedded in the center of the magnets, enabling bilateral communication between connected units (Fig. 5.3(a)). As for the power source, it would be ideal if each unit could run with a built-in battery, but since the electromagnets consume a large amount of electricity, 12 V DC is supplied through a cable.

5.3 Preconditions for Self-assembly Algorithms

In order to realize functions such as self-assembly and self-repair, we need a program, i.e. an algorithm, that specifies the unit behavior. A copy of such a program is loaded on the processor of each unit of the system, and each runs separately and in parallel while communicating with neighboring processors. We should emphasize here that there are many possible conceptual approaches to these algorithms, and that, depending on the conditions we assume, it can be complex or simple. As we discussed in Section 1.3.1, in designing for self-organization, the components are the same in the sense that the functions of one can be carried out by another.

5.3.1 Unit Identifier

If each unit has a unique identifier, e.g. the processor embedded in each unit has a unique ID number such as a product ID, we can define an order among the units based on the IDs. This order can be regarded as a priority order, which is very useful for resolving conflicts among them. The order also ensures that there is one smallest ID number and one largest ID number (This allows symmetry break, which we explain later, that in turn allows centralized control).

However, even when there is no identifier available and all the units are identical both in hardware and software, it is still possible to resolve conflicts or to elect leaders if each unit is capable of generating a unique random number, as we mentioned in Section 4.4.1. Still, in order to finish the leader election within a certain time period, the maximum number of units that the system can contain must be known.

5.3.2 Method and Range of Communication

When units communicate only through individual information exchanges among adjacent units, as in the case of a Fractum, successful communication means that there is a connection and unsuccessful communication means that there is no connection, and therefore a unit can discover its physical connections through communication. In other words, the logical connection of inter-unit communication exactly corresponds to the physical connection of the units. Such information can be used as the beginning state for the self-assembly and self-repair algorithms. It is also possible to get more information by communication between adjacent units (If all the units exchange messages, they update their local information, and then exchange messages again. By this procedure, a unit can discover the connectivity of a neighbor of its neighbor.)

There are other possible communication methods, such as communication over a common bus channel to which all units are connected, or over a wireless network. In these cases, any two units in the system can communicate in a peer-to-peer manner (at the cost of loss due to communication traffic congestion).

However, this kind of communication does not reveal any information on the distance between units (see Section 9.1.3.2).

5.3.3 Spatio-temporal Symmetry Breaking

In considering how to realize self-assembly and self-repair functions, existence of an absolute coordinate system defined in the space makes a significant difference. For instance, self-assembly algorithms differ greatly depending on whether each unit knows its own absolute coordinates (and also its direction) or not. If a unit knows both its current absolute coordinates and its location in the target area expressed in the same coordinates, self-assembly is accomplished simply by moving units outside of the target area towards inside of the area (it is necessary to avoid collisions if a large number of units rush at once). On the other hand, if there is no coordinate information available, the algorithm becomes substantially more difficult, as the units have to determine by themselves their position and orientation in the target configuration.

Some algorithms may require a time origin. In cases where an initial program is executed and then communication between units starts after the power is turned on, the time origin is determined implicitly by turning all the units on at the same time. In the case of staged self-assembly which we explain in Section 5.5, one of the units in an undifferentiated group of units plays the role of providing the origin of assembly. The task of choosing this unit is the same as that of the leader election, and the moment the leader is determined is the time origin of the process of self-assembly. The procedure of breaking the spatio-temporal homogeneity like this is called *symmetry breaking*[3]. It is also a useful method to completely delegate the power of symmetry breaking to a particular unit (e.g. the unit with the largest ID number, or the one chosen as the leader).

As we have seen, there are various possible preconditions, and depending on them the structure of the algorithm greatly differs. In the rest of this chapter, we discuss two such algorithms, which have the following preconditions:

- Algorithm (I) for self-assembly assumes

  (1) No unit identifier
  (2) Direct communication only between adjacent units
  (3) No symmetry breaking

- Algorithm (II) for staged self-assembly and self-repair assumes
  (1), (2) the same as Algorithm (I)
  (3') Symmetry breaking given a priori

The notable feature of these algorithms is that they strictly adhere to the definition of distributed architecture: that the units are perfectly the same, that no coordinate

---

[3] In organisms, symmetry break happens at the moment of fertilization. The process of biological development begins from the moment a sperm binds to an ovum, and the direction of the plane of cell division is determined depending on the position where the sperm binds.

information is available in advance, and that the only communication is between adjacent units. The algorithms (I) and (II) only differ as to whether symmetry is broken a priori or not. It is interesting to see that such a small difference between the two algorithms generates a great difference in the functions they can offer.

5.4 Algorithm (I) for Self-assembly

A feature of a homogeneous mechanical system consisting of the same units is that any unit can be used in any location in the system. In other words, each unit should have the potential to serve as any part of the system. Therefore, all the units must be governed by the same software, and this software must contain information of some sort regarding the target configuration. The *self-assembly* problem is that of transforming the units in a random configuration into the target configuration under such assumptions [1].

[Self-assembly Problem]
Consider a collection of units in an arbitrary configuration, though it is required that all the units are connected to at least one other unit, as shown in Fig. 5.5. Each unit is given information on the target configuration in advance, but it does not know its position among the rest of the units. Each unit can discover its local connections only by communicating with adjacent units to which it is connected. Given such constraints, how can each unit be driven so as to form the target configuration?[4]

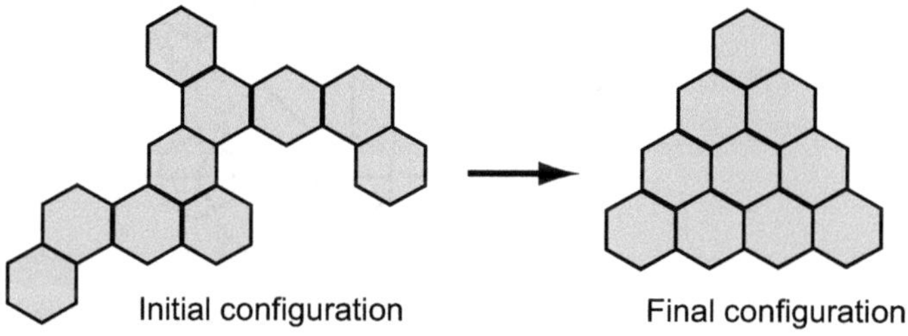

Fig. 5.5 Self-assembly problem

5.4.1 Description of the Target Configuration

Each unit can discover its local connectivity only through neighbor-to-neighbor communication. Thus, we need a way to describe the target configuration by listing the local connections of the units.

[4] Note that since we are considering transformations from an arbitrary configuration to the target configuration, there is no spatial symmetry breaking except for using randomized numbers in Algorithm (I).

5.4.1.1 Connection Type

There are twelve types of possible connectivity that one unit can have, if rotations and mirror images are ignored; these are shown in Fig. 5.6. We call these the *connection types*. The dot in the center of a hexagon denotes a unit and a line from the dot to one of the sides indicates that this unit is connected to the unit in that direction.

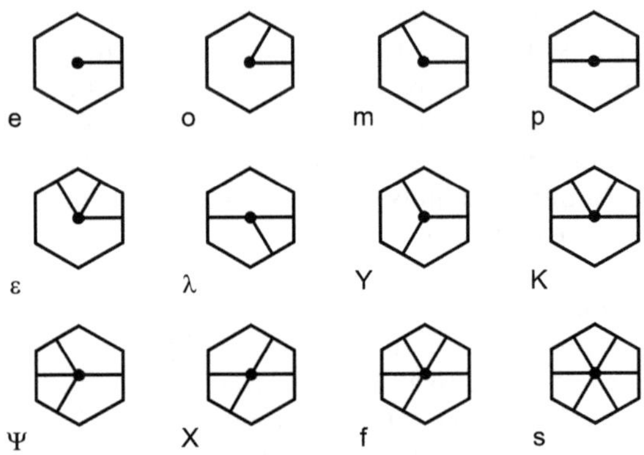

Fig. 5.6 Connection types of Fracta

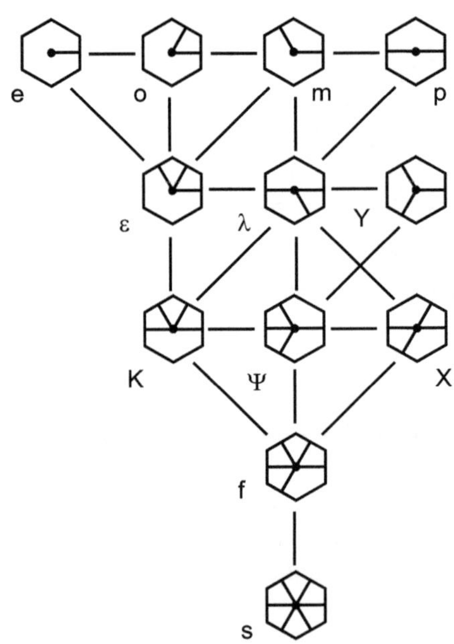

Fig. 5.7 Type transition diagram (distance between types)

5.4.1.2 Distance between Connection Types

The distance between different connection types is defined by introducing a *type transition diagram* (Fig.5.7). The nodes of this diagram are connection types. Two types in the diagram are connected by a link, if a unit in one type can change into the other type after a single step movement (30 degree rotation). Distance between two types is defined as the number of links on the shortest path between them on this diagram. For example, the distance between type e and type X is three. The distance calculated in this way gives only the lowest limit, and it reflects the minimum amount of effort required for transition from one type to another.

5.4.1.3 Description of the Target Configuration Using Connection Types

Each unit knows its own connection type, and it can find out the connection types of the units in its neighborhood by communication. Given a unit, we call the list of the connection types of the adjacent units the *adjacent type list.*

As an example, let us consider the case where ten units are arranged in an equilateral triangle as in Fig. 5.8. This configuration contains three different connection types; o (corners), K (side edges), and s (center). In this case, the adjacent type list of the type o unit is {-, -, -, -, K, K}, that of the type K unit is {-, -, o, K, K, s}, and that of the type s unit is {K, K, K, K, K, K}. Here, the units are listed in the order shown in Fig. 5.6 (e, o, m, p, ε, λ, Y, ...)[5], and the symbols "-" placed at the beginning of a list indicate absence of connections, so that all lists have the same number of elements.

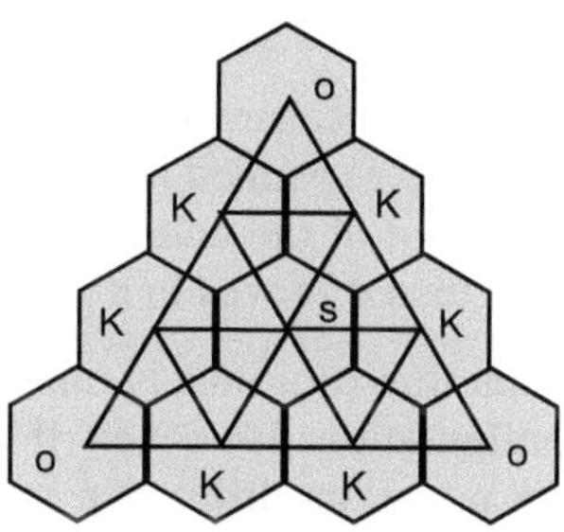

Fig. 5.8 Example of a target configuration description

Now we are ready to describe the target configuration using local connectivity relations only. Let us call the types contained in the target configuration *target types*; we describe the target configuration using pairs consisting of a target type and its adjacent type list. We call the combination of a target type and its adjacent type list a *statement.* Repetitions of the same statements are omitted in a

[5] It is also possible to list the types in the order going around the unit. Although this ordering contains more information on the connection configuration, we chose this order because it simplifies computing the difference, which we discuss in the next section.

description. This means that there are the following three statements for the shape in Fig. 5.8:

o{-, -, -, -, K, K}.
K{-, -, o, K, K, s}.
s{K, K, K, K, K, K}.

When the target configuration is more complex, or it consists of a larger number of units, the description may be longer. Given a configuration, the description is uniquely determined, but there may be multiple configurations that correspond to a given description. For example, since even if type λ is flipped over, the description is unchanged, so the same description may generate several different configurations, as in chemical isomers.

5.4.2 *Strategy for Self-assembly*

Based on the representation of the target configuration only using local information, we next consider how to build the target configuration. Our strategy is the following:

1. In order to evaluate quantitatively the difference between the current configuration and the target configuration, each unit calculates its *difference measure* (a unit whose difference measure is zero is already a part of the target configuration).
2. Check whether each unit is of the *movable type*. A movable unit can move as long as the motion does not cause any part of the collection of units to disconnect.
3. Estimate the average difference measure by an equation simulating the diffusion of each unit (*diffusion field*, explained later). If the difference measure of a particular unit is higher than that of the units around it and it is movable, decision to activate is made, and then it moves in a random direction.

With this strategy, since the motion of a unit is random, it is not apparent that it will approach the target configuration. However, if it does approach the target configuration as a result of a move, its difference measure is decreased and that makes the next movement of this unit more difficult. Since all the units behave in this way, the entire collection of units gradually approaches the target configuration. In the rest of this section we give further details of this strategy.

5.4.2.1 Difference Measure

In order to form the target configuration with distributed units, each unit needs to know which part of the target configuration is closest to its current local situation. Therefore, for a given unit, to indicate the similarity between its current local situation (its own connection type and its adjacent type list) and the statements describing the target configuration, we define its "*difference measure*" Δ as follows:

$$\Delta(i) = \min_j \left[d(\, type(i),\ \ type^{\mathrm{d}}(j)\,) + \Sigma_k\, d(\, ntype(i, k), ntype^{\mathrm{d}}(j, k)\,) \right],$$

where

$d(a, b)$: the distance between type a and type b
$type(i)$: the current type of the ith Fractum
$ntype(i, k)$: the kth type in the current adjacent type list of the ith Fractum
$type^d(j)$: the type of the jth target statement
$ntype^d(j, k)$: the kth type in the adjacent type list of the jth target statement.

This is evaluated at every time step. Specifically, for each statement in the target description, the distance between the unit's current type and the type of the statement (the first term) is calculated, and then the unit compares its own adjacent type list with the adjacent type list of the statement to calculate the distance between each pair of corresponding units (the second term). Then, for each statement these distances are summed, and the minimum distance gives the difference measure Δ of the unit. If a unit becomes a part of the target configuration, its difference measure becomes zero. When the target configuration is completed, the difference measure becomes zero for all units.

5.4.2.2 Movable Type

The ability of a unit to move is determined as follows. In situations where many units are connected to each other, not all the units are movable. A unit's movability depends on several factors: one is the hardware specifications of the Fractum which determine whether it can move without collision (for example, a type "s" unit cannot move because all of its slots are filled. Similarly, any unit with four or more connections cannot move.) There is also a constraint due to the actuator specification of Fractum that each unit can generate only enough torque to move itself. Moreover, units of type "m" and "p" are not allowed to move because parts of the unit will become disconnected if units of those types move. For the above reasons, the connection types "e", "o", and "ε" are the only movable types.

If we give priority to the units with larger difference measure to be moved first, one may think that the target configuration will be reached eventually. However, since units are only capable of local communication, it is not possible to examine the whole system and select the unit with maximum difference measure. Instead, we choose to move those units that are movable and have differences measure relatively larger than the local average. The direction of the move is randomly determined. This is due to the fact that it is difficult to determine locally which direction is the best to move in to facilitate the formation of the target configuration. Still, if a random move of a unit makes the configuration closer to the target one (the difference is decreased), the priority of that unit is decreased and the unit is less likely to move in the next step. This also prevents regression of the formation process. With this strategy, units to be moved are chosen locally, and multiple units are allowed to move simultaneously in a large system.

5.4.2.3 Diffusion Field

In order to implement the strategy described above, it is necessary to estimate for each unit the local average of the difference measure. This is computed using the diffusion field. The problem of estimating the average value of a certain parameter with continuous values through local communication is solved by the following diffusion equation:

$$\mathrm{d}x(i)/\mathrm{d}t = K \sum_{j=1}^{n(i)} (x(j) - x(i))/n(i)$$

where

$x(i)$: diffusion variable of the ith Fractum
$n(i)$: the number of fracta that connects to the ith Fractum
K : diffusion coefficient.

The initial value of the variable $x(i)$ is set to be the difference measure of the unit, and is changed according to this equation for a sufficient amount of time. The value of x then can be used as an estimate of the average of the local differences. At every instance when the connection configuration of the unit changes, the value of x is reset to its difference measure at that time.

5.4.2.4 Activation Criteria

To determine whether the difference of a unit is larger than the local average, we use the following criterion.

$$Gx(i) < \Delta(i)$$

where G is an activation constant which is greater than 1. If this is satisfied, and the unit is movable, it moves one step immediately (if it is not movable, nothing happens). This process is called the *activation* of a unit. It is clear that a unit whose difference is zero is never activated, and units with relatively large difference will be activated when x decreases due to the diffusion process.

However, every time a unit moves and the x values of its own and neighbor units are updated, new differences will be put into x, and the total sum of $x(i)$ tends to grow rather than remain constant. This is because a movable unit with small $x(i)$ is activated and possibly will get a larger value after its movement.

To avoid this, we introduce a leak only at movable units.

$$\mathrm{d}x(i)/\mathrm{d}t = K \sum_{j=1}^{n(i)} (x(j) - x(i))/n(i) - L$$

where L is the leak constant[6]. This makes movable units more likely to move.

[6] In order to keep x positive, when updated x become negative, 0 is substituted to it.

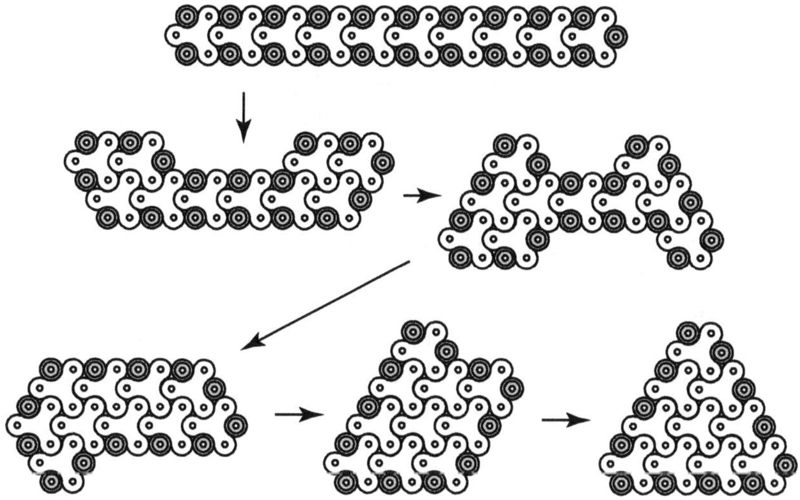

Fig. 5.9 Simulation of self-assembly

Although the average of difference measures decreases to zero as the formation process proceeds, since activation is decided by comparison with this average, units with relatively large difference are sure to be activated at some time. In this way, until the formation is complete, units continue to be activated one after another. When the formation is complete, the difference of all the units converge to zero, and therefore no further move occurs.

5.4.3 Simulations and Experiments

Computer simulations were made to evaluate the validity of Algorithm (I). We used an initial configuration where the units are aligned in one row.

1000 simulations were executed aiming to form a triangle from ten units, and the success rate that the target configuration was reached within 2000 steps was 96.8%. A sample of a successful case is shown in Fig. 5.9. In the unsuccessful cases the units got caught in a deadlock (a state with no movable units) (Fig. 5.10).

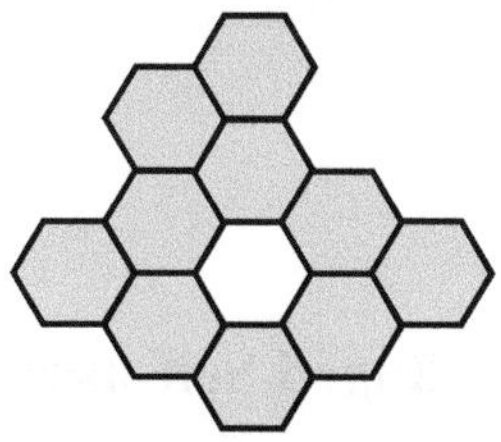

Fig. 5.10 Example of a deadlock. Once a hole is formed, units around the hole can never be moved again, because this configuration includes no movable units

Fig. 5.11 shows the changes in the totals of difference measure and diffusion variables for successful cases. Although the plots have an overall direction, they demonstrate that the search is basically random. Consequently, when the number of units increases, the convergence rate drops drastically. For example, if we increase the size of the triangle by one row (15 units), the success rate within 4000 steps dropped to 73.4%. In the unsuccessful cases, the units either continued to move without converging, or were caught in deadlock.

Fig. 5.12 shows an experiment in which Algorithm (I) was executed with real units [2]. The cables are only for supplying power. Since the same algorithm as the simulations was used, the units moved basically in the same way. Due to physical constraints, however, changing of connections failed sometimes, resulting in unsuccessful assembly when the number of steps was large. This experiment served its purpose of validating the principles.

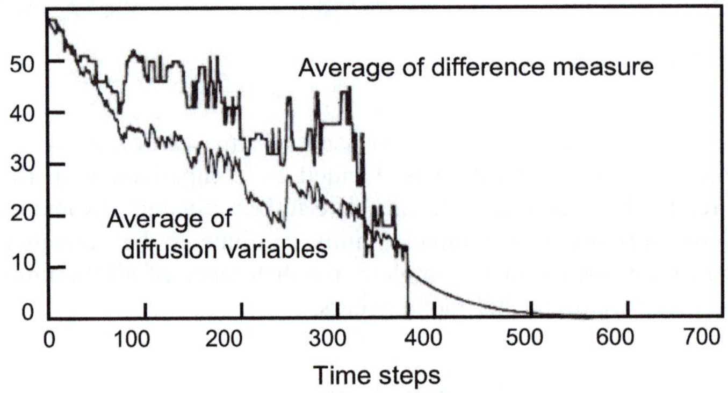

Fig. 5.11 Graph of changes in difference and diffusion

Fig. 5.12 Self-assembly experiment

5.5 Algorithm (II) for Staged Self-assembly and Self-repair

Though Algorithm (I) is successful in the above self-assembly, its effectiveness is limited; the number of units in the system needs to be exactly the same of that of the target, and the success rate goes down for a larger target or for an asymmetric target.

In the rest of this chapter, we describe Algorithm (II), which tries to resolve these issues by introducing a coordinate system over the units. By setting a point of origin in the collection of units, which allows specification of the position of finished assembly and the axis of symmetry in advance, it is possible to reduce unnecessary moves (searching cost) greatly, hence enabling assembly of larger scale and more complex configurations [3].

The outline of Algorithm (II) is as follows: first, select one unit from the group and set it as the origin of assembly. We call this unit the *kernel*. This process of selecting a kernel unit is the symmetry breaking which we discussed in 5.3.3. The difference between a kernel and other units is simply that a kernel knows that it has been chosen from among the others, otherwise, all the units execute the same program. Next, the kernel unit forms a "virtual" network with adjacent units, based on an assembly blueprint describing every stage of assembly. The blueprint which we call a *description matrix* is carried by every unit. This matrix describes how this virtual network shall develop in each stage.

When the kernel finishes building the first stage of the virtual network, those units which were incorporated in the first stage layer start expanding the network further by taking in adjacent units outside the network to build the second stage of the virtual network. By repeating this process, the third, fourth, and further stages are built, until the entire body of units coincides with the target shape. We call this algorithm the *onion method* because of the way the network expands from a single kernel by repeatedly growing a layer to surround the previous layer (Fig. 5.13).

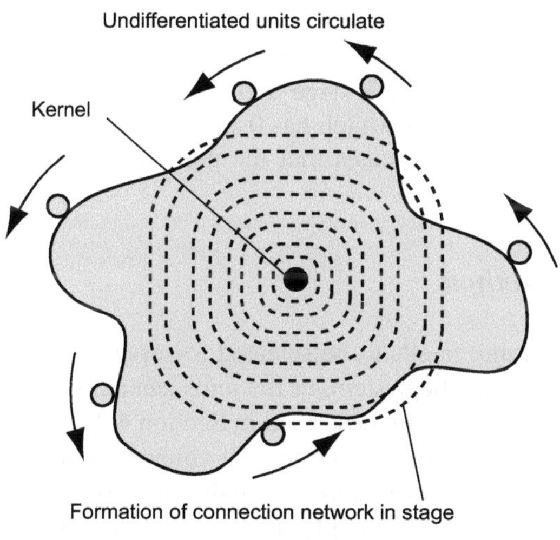

Fig. 5.13 Concept of onion method

5.5.1 Logical Type and Description Matrix

In order to expand the network, there must be units in the area where the network is expanding. If the entire space is filled with units as in the case of cellular automata, the task required is simply to form new connections for the network. Here, though, there are only a finite number of units, and we need to supply units to the area. This is done by circulation of undifferentiated units. Also, operations are possible only at positions where units exist. In order to deal with such a situation, we introduce the new notions of *logical connection* and *logical type*. Correspondingly, we now call the connection types we used in Algorithm (I) *physical types*. While the physical type of a unit is determined by its physical connections, in Algorithm (II), each unit is regarded to have virtual connections independent of its physical type.

In the following section, we assume that the space is filled with an infinite number of units, and consider a method to form a network of logical connections, starting from a kernel chosen from among those units. In order to adapt this method to the case where the number of units is finite; it is only required that units that are included in the network of logical connections stay as they are, while those that are not included move about freely. In this way, even parts of the logical connection network that grow into areas where there are no units will be covered over by units eventually (See the last part of 5.2.2).

Next, we explain the description matrix. Each unit has an assembly blueprint called a *description matrix*, which contains information on which logical type a unit (at a certain stage of assembly and at a certain position within the system) should have. Each unit uses two parameters, *stage* and *location index*, to scan the description matrix to determine its logical type. A description matrix is a lower triangular matrix as in Fig. 5.14, where the horizontal axis indicates the location index and vertical axis indicates the stage.

The initial values are 0 for the stage and −1 for the location index for all the units except for the kernel unit, which has 0 for the stage and 0 for the location index. A dash in the matrix indicates that the unit should keep its current logical type.

5.5.2 Onion Method

To start assembly, a unit needs to be selected to serve as the kernel. This choice can be made from outside the system, or the units can autonomously select one using the leader election algorithm described in Section 4.4.1. The unit chosen as the kernel immediately starts building the logical connection network as specified in the description matrix. What the kernel actually does is to send out messages to undifferentiated units to which it is physically connected (an undifferentiated unit is one whose stage and location index are both undetermined) requesting them to join in the logical connection network. The message contains the stage and the location index of the sender so that a unit which receives it can determine the logical type which the receiver must adopt from the description matrix. However, the description matrix does not indicate the directional orientation of that logical type. Therefore, the unit which receives the message selects a tentative direction of its logical type at random. After all the units that received messages put their logical type in random direction, if all the units are satisfied[7], the logical network for that stage is complete.

How to recognize when a stage has been completed is a significant problem. This is due to the fact that the area of a logical network may be much bigger than the neighborhood of each unit. It is necessary to detect completion before going on

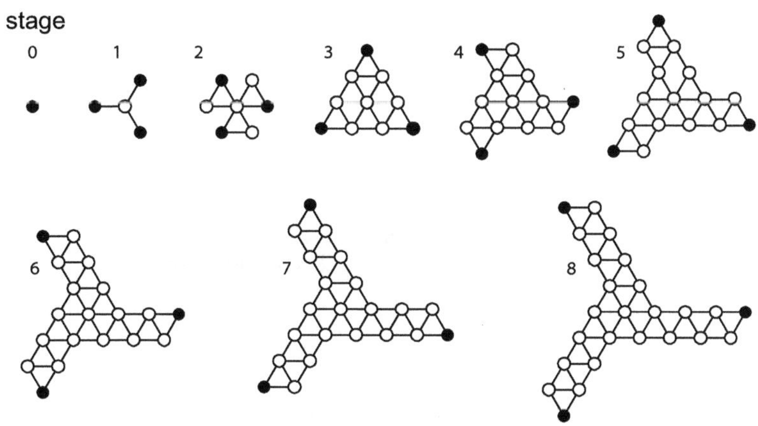

stage (s) \ location index (i)	0	1	2	3	4	5	6	7	8
0	n								
1	Y	e							
2	s	o	o						
3	—	K	K	o					
4	—	—	K	e	o				
5	—	—	—	K	e	o			
6	—	—	—	—	K	e	o		
7	—	—	—	—	—	K	e	o	
8	—	—	—	—	—	—	K	e	o

Fig. 5.14 Assembly stages and description matrix

[7] A unit is *satisfied* when all the units to which it sends requests through its logical connections do the same and send requests back to it. When all the units in a stage are satisfied, the logical network for that stage is *complete*.

to the next stage. For this purpose, we use the following *minimal completeness propagation algorithm*, which uses the properties of a diffusion system discussed in Section 4.3.1:

[Minimal Completeness Propagation Algorithm]
Each unit is assumed to have a positive integer value called *completeness* in this algorithm. Also, an integer threshold value R is decided upon in accordance with the size of the system. At a given time, the completeness of a non-satisfied unit is set to zero. In the case of a unit that is satisfied, its completeness in the next time step is calculated by detecting the completeness of all the adjacent units and adding 1 to the lowest of these completeness values (Fig. 5.15).

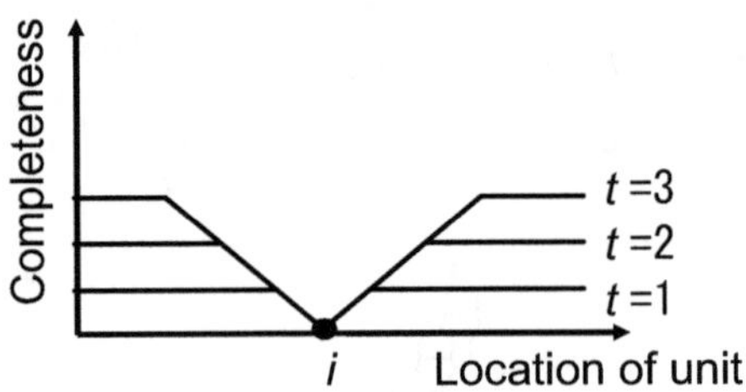

(a) Time development of completeness distribution when all except for the i-th unit are satisfied at time 0

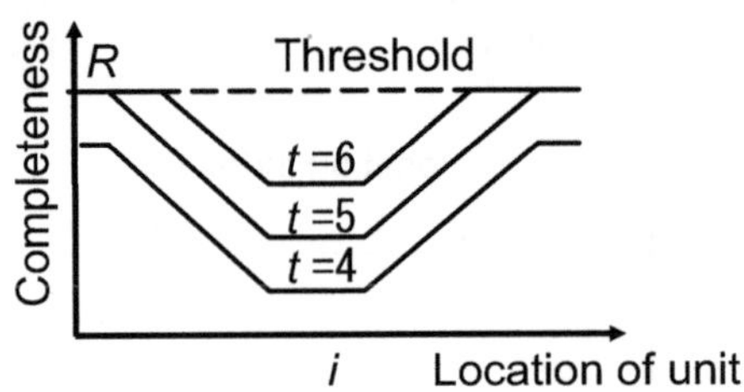

(b) Time development of completeness distribution when the i-th unit also becomes satisfied at time 4

Fig. 5.15 Minimal completeness propagation algorithm

In this way, once the network for a particular stage is completed, the completeness of every unit increases with every time step, and eventually they all pass the threshold R. Therefore, given a unit, when all the units within the circle with a radius R around the unit are satisfied, the unit switches to the next stage. Note that all the units do not progress to the next stage at one time but rather progress unit by unit as they are completed. This process is repeated until the last stage described in the description matrix.

As we mentioned earlier, as the assembly process progresses, there comes a time when the logical network must expand farther than the boundary of the physical units. If undifferentiated units can always be supplied to the points where the network is expanding, the network can keep growing as long as there are units available and eventually take on the physical shape of the target configuration.

In a two-dimensional machine, whatever shape the system takes, its perimeter is a single closed curve. Using this property, it is possible to supply units easily.

[Unit Supply by Circulation]
A unit is either movable or non-movable depending on its physical type. Just as in the case of the Algorithm (I), we categorize the connection types "e", "o", and "e" as movable. If we make a movable and undifferentiated unit have some probability of rotating one step to the left, any undifferentiated unit located on the perimeter of the whole structure can circulate counterclockwise around the logical network (Fig. 5.13).

5.5.3 Simulation of Self-assembly (Algorithm II)

When applying this algorithm, each unit should follow the steps below:

1. Select a kernel
2. Determine its stage and its location index, determine its logical type
3. Detect the direction of its logical type
4. Transmit its completeness and judge whether the stage should be changed
5. Move undifferentiated units

After step 1, each unit simply repeats steps 2 to 5. The execution of this algorithm is completely distributed and parallel. Moreover, all the units have identical programs and data in initial state. The data exchanged by the units include random numbers for leader election, physical types, stages, requests for logical connections, and completeness, all of which are expressed as integers.

We conducted a computer simulation to evaluate the algorithm (Fig. 5.16). In this diagram, the units are indicated with numbers, and the number indicates the stage of that unit. 0 indicates an undifferentiated unit. In the initial state, the unit in the middle is the kernel. The process of leader election is omitted and so the kernel is at the center from the beginning. Lines between units indicate logical connections. Using the description matrix given in Fig. 5.14, we ran 1000 simulations from the same initial state, and in all cases the assembly was successfully completed (since the algorithm uses random numbers, the simulations are different every time even with the same initial configuration). The average time required to reach the final stage was 590 steps.

Next, we ran simulations with a more complex target configuration of 91 units (5 of these kept as spare parts). The result is shown in Fig. 5.17. Out of 1000 simulations we ran, in 979 cases the assembly was completed within 5000 steps. In these cases, the last stage (25th stage) was reached after 2600 steps on average. In the unsuccessful cases, either the assembly was not complete after 5000 steps, or the system fell into a deadlock due to a hole formed during the assembly. We also experimented with several other target configurations, and in most cases, the assembly was successful with high probabilities.

5.5.4 *Simulation of Self-repair (Algorithm (II))*

Algorithm (II) can be extended easily into a self-repair algorithm, because of its hierarchical structure. In the rest of this chapter, we discuss self-repair of a system, that is, the way that a system repairs a failure of any of its units [3].

Since it is difficult to take all possible failures into consideration, we limit the type of failures we consider here to random multiple units becoming disconnected from the system. We allow the possibility that any unit at any position becomes disconnected at any time, but we assume that all the other units function normally. The system should detect the missing units, and replace them with spare units (this might require disassembling of already built parts, retrieving undifferentiated units, and repairing with these).

These assumptions are realized easily by retracing the steps of assembly. That is, self-repair can be realized by stopping the process of assembly still underway, retrogressing to the stage at which the system is certain to be free from failures, and restarting the assembly from that point.

5.5.4.1 Detection of the Loss

Units that are already differentiated and part of the logical network are in charge of detecting the loss of units. When the signals are lost from a connection arm that is supposed to be logically connected to another unit, that unit is considered to be lost.

5.5.4.2 Retrogression Signal

Retrogression to earlier stages is started when the unit which detected the loss sends retrogression signals to units around it. The retrogression signal contains information of the stage to which the system should return (retrogression level). The level (depth) of retrogression is determined using the location index of the lost unit. This is for the purpose of keeping the logical network of the highest stage which is still intact in the remaining system.

5.5.4.3 Retrogression of the Stage

In the case where the unit that detected the loss sends out a retrogression signal to return to level n, the entire system has to be set back at least to the stage n. Fig. 5.18 will help us explain this process. In order to set back the stage, it suffices to set back those units in the system whose stage is higher than n (corresponding to the regions B) to the stage n. Those units in the region A become undifferentiated again.

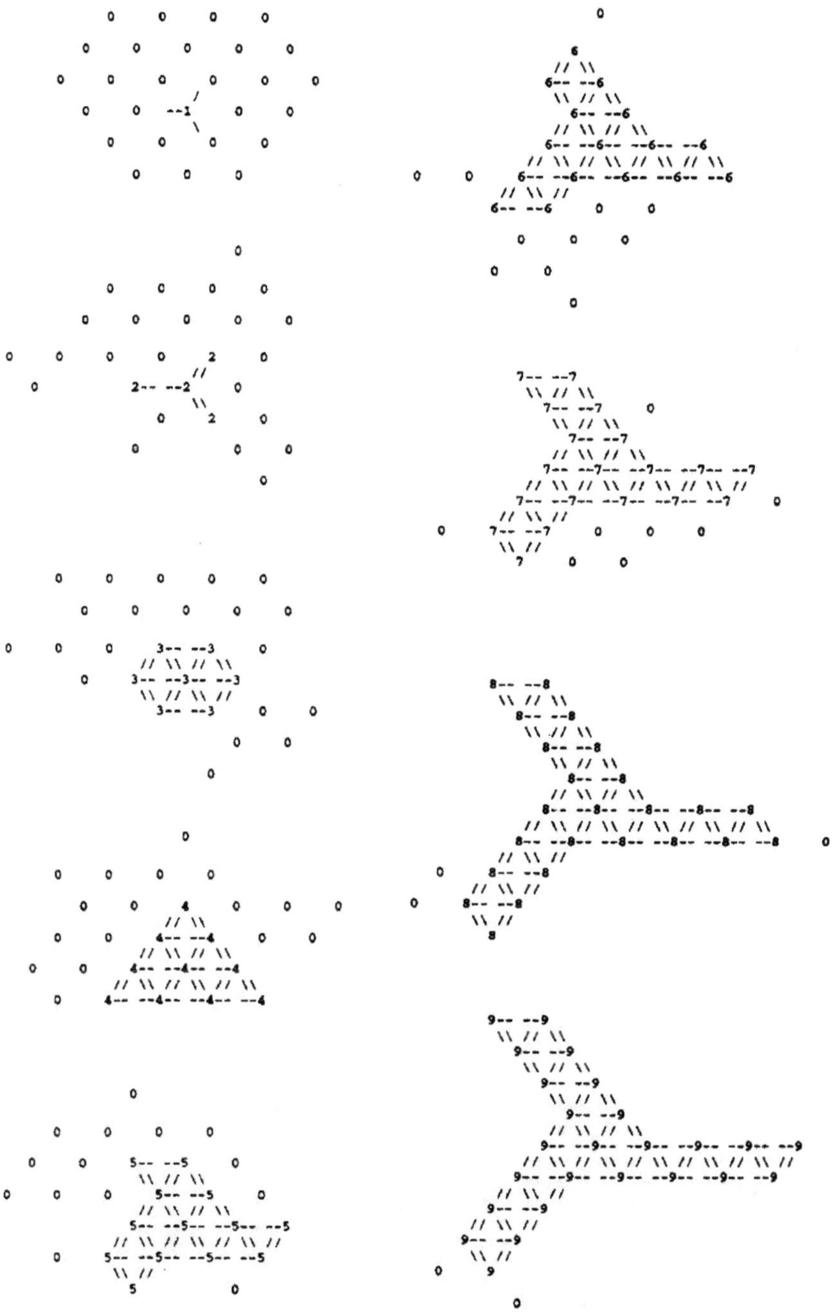

Fig. 5.16 Assembly simulation using the description matrix in Fig. 5.14

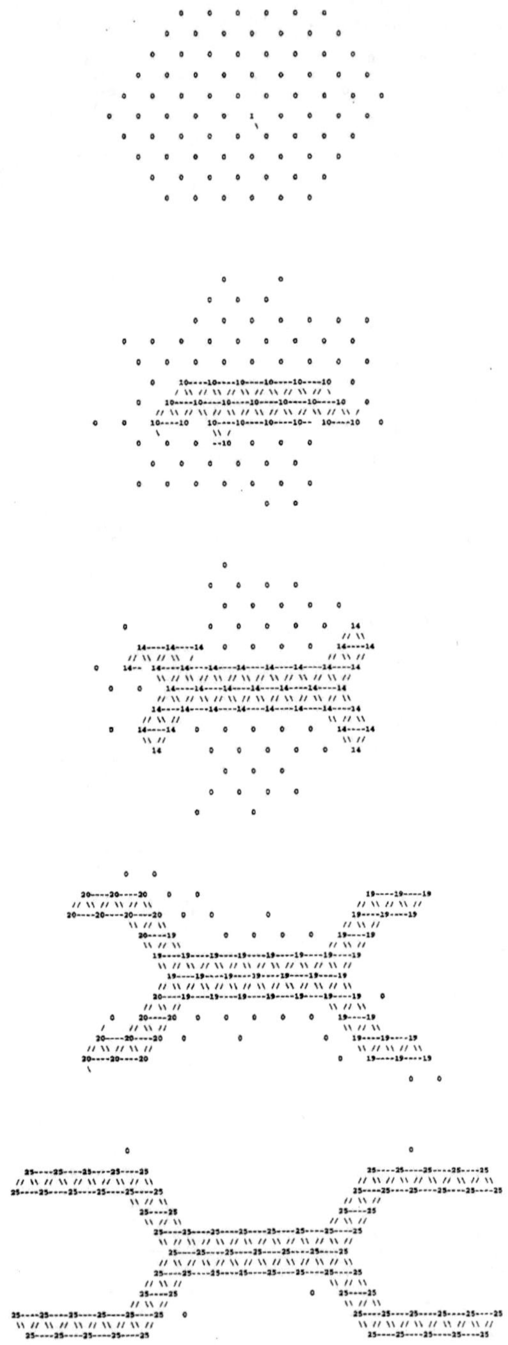

Fig. 5.17 Self-assembly of complex configurations

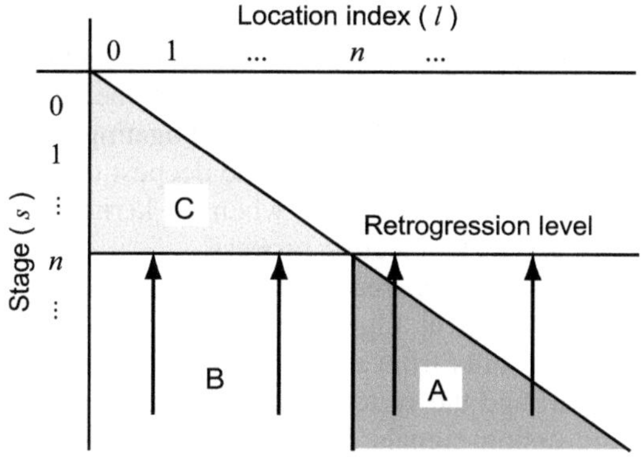

Fig. 5.18 Self-repair using a description matrix

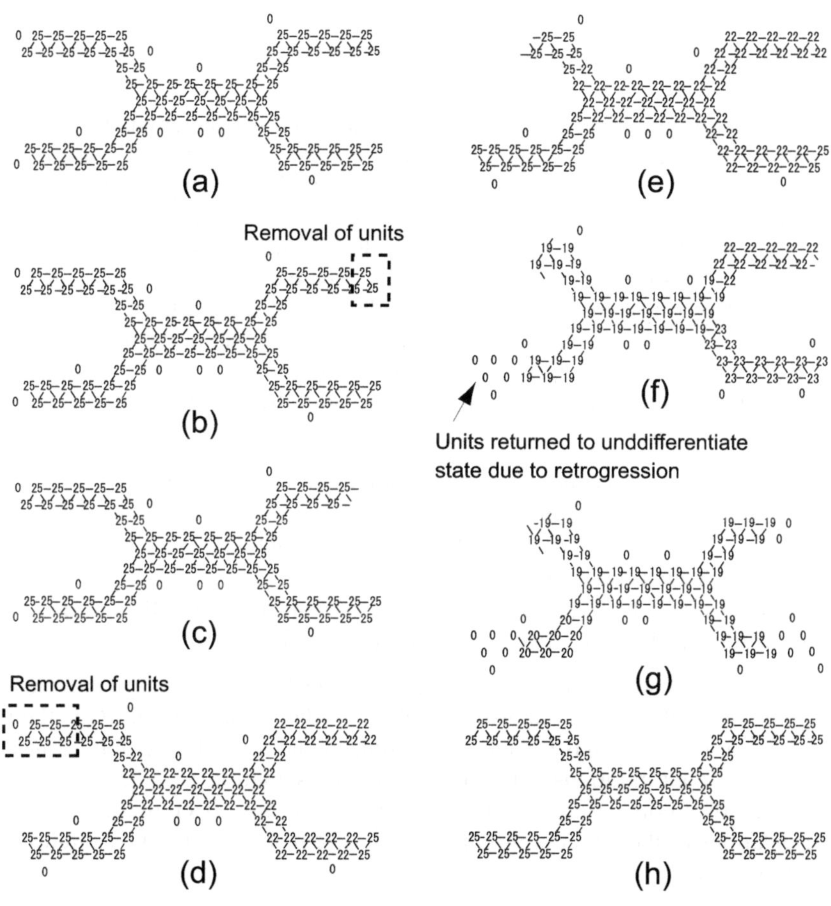

Fig. 5.19 Simulation of self-repair after unit removal.

In this method, the unit that sends out a retrogression signal can resume its assembly immediately. Thus, the process of self-assembly and that of self-repair can be executed concurrently. When several units are lost one after another, retrogression signals indicating different levels will be propagating through the system. In this case, the system will choose to return to the deepest of these retrogression levels (the smallest stage number). However, when the kernel has been lost, the system has to restart from the leader election process.

Fig. 5.19 shows results of a self-repair simulation. Several units were removed from the configuration shown in Fig. 5.17 and the self-repair process was executed. First, we removed units from a completed configuration. Units "0" are the spare units circulating around the system. Diagrams (a) to (f) in Fig. 5.19 show the transmission of retrogression signals. The whole system tries to get back to the stage corresponding to location index 19, with the remaining units (Fig. 5.19(g)). After the retrogression is complete, the process of reassembling is exactly the same as the normal assembly, and the original configuration is regained.

5.6 Cellular Automata Model

The Fractum system driven by the two algorithms in this chapter can be viewed as cellular automata. As cellular automata, the units are supposed to be scattered on a hexagonal lattice space. A cell (lattice node) is live if it is occupied by a unit, otherwise it is dead. Each live cell decides its motion based on its state and its neighbors' states according to a set of transition rules. Rotation of a unit corresponds to simultaneous transitions of two adjacent cells, where a live cell becomes dead and a dead cell live (this must be consistent with the real unit). Strictly speaking, to homologize Algorithm (I) to be action of a cellular automaton, the diffusion variable x in Section 5.4.2.3 need to be represented by a discrete number.

Such cellular automata are a little different from the true cellular automata described in Section 4.3, as

- A dead cell has no capacity for processing, and state transition of a dead cell needs to be processed by an adjacent live cell.
- The total number of live cells needs to be kept constant.
- All the live cells need to be connected.
- The system may be asynchronous, i.e., transition of each cell may occur any time as long as the above two conditions are satisfied.

Similar studies of asynchronous cellular automata have been made not only for hexagonal units like Fracta but also for cubic or other lattice systems (see the next chapter). Various tasks have been considered, including self-assembly, transition between two configurations, the whole structure's motion as a fluid flow, and reaching a given point. Various settings for problems have been used, for example representations of the target configuration varying from a local representation to a precise global one.

References

[1] Murata, S., et al.: Self-Assembling Machine. In: Proc. IEEE Int. Conf. Robot. Autom., vol. 1, pp. 441–448 (1994)
[2] Yoshida, E., et al.: An Experimental Study on a Self-repairing Modular Machine. Robot. Auton. Syst. 29(1), 79–89 (1999)
[3] Tomit, K., et al.: A Self-Assembly and Self-Repair Method for a Distributed Mechanical System. IEEE Trans. Robot. Autom. 15(6), 1035–1045 (1999)

Chapter 6
Prototypes of Self-Organizing Robots

Abstract. In Chapter 3, we introduced a line of research initiated by von Neumann and Penrose around 1960. The ultimate goal of their work was artificial self-reproducing systems. In Chapter 5, we explained our mechanical unit called Fractum, whose goal was self-assembly and self-repair. In fact, the Fractum module was one of the first systems proposed by various researchers after microprocessors became commercially available. In this chapter, in order to help the reader get an overall picture of the research on so-called modular robots, we make a classification of modular robots, identify typical problems in their design, and then give an overview of some representative systems that have been developed.

6.1 Classes of Modular Robots

In the early model-oriented research of von Neumann's automata and Penrose's building block models, the system structure itself was of chief interest. Following these seminal studies, the research on self-organizing mechanical systems gradually shifted to machines that actually change shape and move around, i.e. robots (See Fig. 6.5). In our discussion of the Fractum (Chapter 5), which also belongs to this line of research, we used the term "unit". The self-organizing mechanical systems we discuss next are composed of such units and are intended to be like robots, not merely machines, and therefore we shall refer to them as *modular robots*.

There is a wide spectrum of modular robots, from simple assemblies of modules to those capable of advanced self-reconfiguration. We start by briefly presenting a classification of modular robots.

Class 1: Modular Robots with Fixed Configuration
A modular robot in this class consists of several modules, each of which has certain level of independence. When this robot contains multiple modules that are identical, the design and maintenance is simplified. However, the connection topology between modules (*configuration*) is fixed and cannot be changed.

Class 2: Manually Reconfigurable Modular Robots
The connection topology among modules of a robot in this class can be changed manually by a human operator. Block toys, like LEGO MINDSTORMS®, belong to this class. Designing robots in this class requires more elaboration because the modules should be equipped with standardized connectors and the software

S. Murata and H. Kurokawa: Self-Organizing Robots, STAR 77, pp. 105–130.
springerlink.com

running the modules should implement standardized protocol. However, once the modules are built, they operate no matter how they are connected, so that various functions can be easily realized using them.

Class 3: Self-reconfigurable Modular Robots
A robot in this class consists of modules that are equipped with the actuators, sensors and control circuits which are necessary for changing connections, and are capable of automatic reconfiguration without human involvement. Fractum which we discussed in Chapter 5 belongs to this class.

Class 4: Self-replicable Modular Robots
A robot in this class is capable of assembling modules that are scattered around into a duplicate of its own configuration. The kinematic model of von Neumann we discussed in Section 3.1.3 belongs to this class.

Of the classes described above, those robots in Class 3 and Class 4 are considered to be self-organizing mechanical systems.

6.2 Lattice-Type and Chain-Type

We can also categorize modular robots into two types: lattice-type and chain-type, independently of the classification we gave above (Fig. 6.1). In this section we give the definition of these types and consider their advantages and disadvantages.

Modules of *lattice-type* form a crystalline structure as in the case of Fractum. A crystalline structure is formed in a space when many modules gather to form a periodic structure retaining a specific geometric symmetry. The procedure of module reconfiguration to form this structure can be determined easily, because a lattice-type module can only move to neighboring points in the lattice. On the other hand, lattice-type modules require many connectors and degrees of freedom depending on the geometric symmetry. One way to reduce the number of actuators is to join two modules together. Although the module becomes less symmetric, it makes the mechanisms simpler because connector mechanisms for these joints become unnecessary.

Modules of *chain-type* can be considered as segments of a multi-joint robot which can be taken apart at the joints and reattached with connector mechanisms. It is easy to build multi-leg multi-arm robots with chain-type modules by introducing a few branching modules. Chain-type modules have advantages over the lattice-type in that symmetry requirements are relaxed and the same functionality can be realized with fewer actuators and connectors. They are also suited for robot-like flexible motions because of their multiple degrees of freedom. On the other hand, reconfiguration of chain-type modules is generally difficult. This is because the positions and angles of modules take only discrete values in lattice type modules, whereas they take continuous values in chain type modules. Therefore, in the latter, positioning connectors to join modules requires measurement of relative

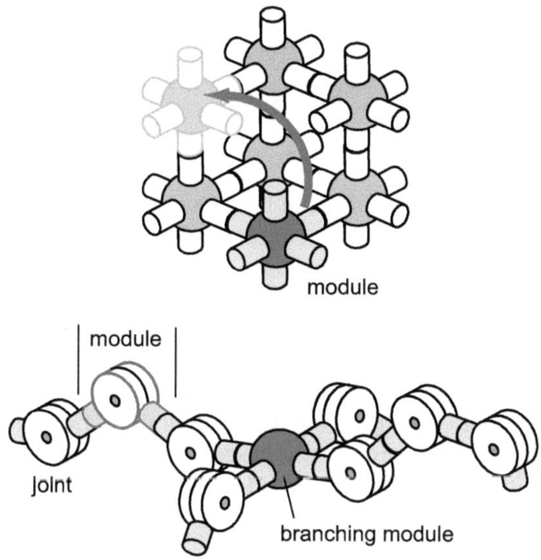

Fig. 6.1 Modular robots of lattice-type and chain-type

positions of the modules and control of many joints including calculation of inverse kinematics.[1]

In addition to these two types, there are truss-type and amorphous-type modules, although few of these have been built. See Section 6.4 for examples.

6.3 Constraints in Hardware Design for Lattice-Type Modules

In designing hardware for lattice-type modular robots, there are several demanding constraints that perplex designers. Designing lattice-type modular robots is like puzzle solving in a sense, because multiple intertwined geometric constraints need to be resolved. It is usually significantly more difficult than designing chain-type ones. In this section, we summarize the constraints which must be considered in lattice-type modular robot design.

6.3.1 Limited Space for Design

A lattice type module has a limited space in which to carry out its functions. The hardware for all necessary functions has to be fitted inside this space. Planar machines (two dimensional machines), like Fractum (Chapter 5) are relatively easy to

[1] There are cases where position alignment is impossible (uncontrollable) due to the number or the direction of joints. See Section 9.3.2.

design because modules can extend in the direction perpendicular to the plane as far as necessary.

6.3.2 Symmetry

The crystalline structure of modules can be generated from the geometric symmetry that the modules have. Designs for most of the mechanical modules that have been made are based on space-filling shapes, as we discussed in Section 5.2. In the case of two dimensional machines, tessellation with one symmetric shape is possible with the three regular polygons shown in Fig. 5.2: triangles, squares, and regular hexagons. For three dimensional space, again there are three possible space-filling polyhedra that are symmetric: cubes, truncated octahedra (Kelvin's fourteen-faced polyhedra) and rhombic dodecahedra[2] (Fig. 6.2). Three dimensional modules developed so far are mostly based on cubes, which have the simplest structures. There are some examples based on rhombic dodecahedra, but none on truncated octahedra.

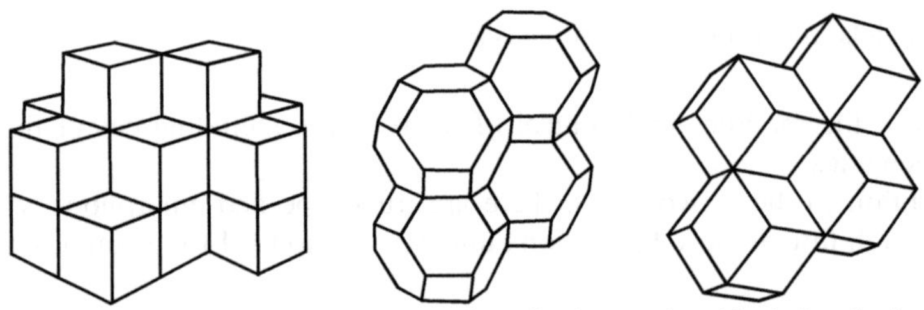

Fig. 6.2 Three dimensional space filling polyhedra

6.3.3 Degrees of Freedom for Mobility

After the module symmetry is determined, the next thing to decide is what kind of mobility to give the modules. Generally speaking, it is enough to consider the degrees of freedom (DOF) necessary for a module at one lattice point to move to a neighboring lattice point. The number of actuators required for a module depends on its required DOF for mobility. In the case of a two dimensional square lattice, for example, there are two possible types of motion: translation and rotation, which allow a module to move to the next lattice point (Fig. 6.3)[3].

---

[2] Hexagonal and triangular prisms allow space filling tessellation but we do not consider them here because they are not fully symmetric (the axes are not equivalent).

[3] Actual implementation of these motions can be done not only in a shape-preserving manner as shown in Fig. 6.3, but also in the ways that modules may be expanded, contracted, or deformed.

Realizing motions like these is not an easy task. In the case of translation, sliding mechanisms have to be installed on the faces of a module. Careful planning of the procedure to build a desired configuration is required, because modules cannot move to a position where they are not supported (See Section 6.4.7). In the case of rotation, mechanisms need to rotate the module about one of its vertices, and also should rotate without colliding with other modules.

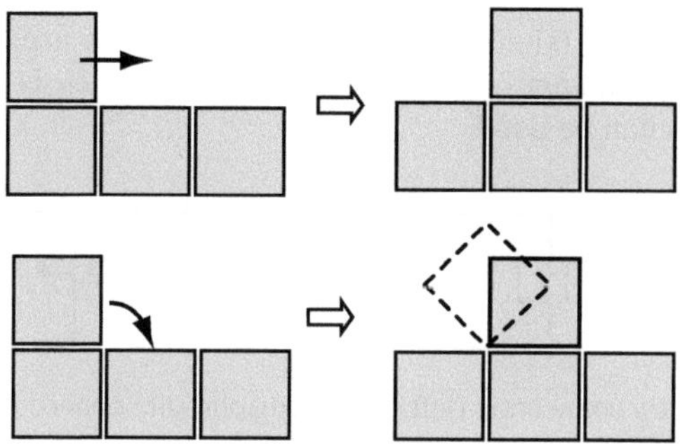

Fig. 6.3 Degrees of freedom in translation and rotation

6.3.4 Connectors (Connection Mechanisms)

One of the most important issues in designing a module is the connector design. The primary function required of a connector is to connect two modules with sufficient strength, but there are more requirements:

1. The connectors must have complementary shapes. Usually, in such a connector pair, one of the pair is the active connector, while the other is passive, and it should be possible to control the connection by manipulating only the active connector. Though connectors may have a self-complementary shape, that is, they may be hermaphrodite (Fig. 6.4), in which case the symmetry of modules is increased, a pair of modules must then have two active connectors, which means not only that more actuators are required but also that both modules must be simultaneously controlled to connect or to disconnect them.
2. The modules must not interfere with each other during their motion. In the case of rotational motion, no other module should be within the radius of rotation of a module (the area that a module moves through when it rotates). In the case of translation, the sliding faces have to be completely flat.
3. Relative position errors and angular errors between the modules should be tolerated to a certain extent.

Energy consumption, information exchange between modules, ease of production and maintenance, etc. should be taken into account as well.

6.3.5 Actuators

The motion and the connection of modules are driven by actuators. The number of actuators depends on the symmetry of the lattice, but in general a high power-to-weight ratio is required. Devices such as DC motors, servomotors, electric magnets, solenoids and SMA (shape memory alloy) are often used, all of which have advantages and disadvantages. When the size of modules is large, hydraulic or pneumatic actuators can be used.

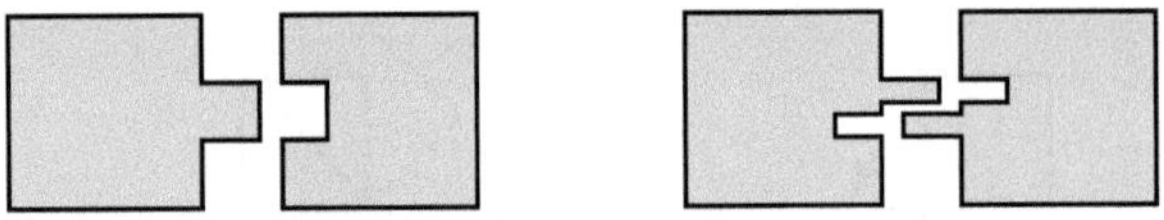

Fig. 6.4 Complementary connectors (left) and hermaphrodite connectors (right)

6.4 Prototypes of Modular Robots

Fig. 6.5 is a rough description of the genealogy of research since von Neumann and Penrose. The lines here help us to understand the relations among them, for example among self-assembly systems using random collisions starting from the self-replicating blocks by Penrose. Similarly, the lineage of lattice-type modular robots starts from CEBOT, which developed into chain-type modular robots such as PolyPod. In this section, we consider several representative prototypes of these lines of development.

6.4.1 CEBOT

For a while after the work by von Neumann and Penrose, this area of research was almost forgotten. As various electronic devices such as microcomputers became widely available in the late 1970s, though, the situation changed suddenly and a variety of research on self-organizing mechanical systems was started. This movement was led by Toshio Fukuda of Nagoya University, who published his work on a dynamically reconfigurable robot system called CEBOT (Cellular Robotic System) in 1985. He proposed a system in which a robot consisting of many modules reconfigures its own structure depending on the task given to it and the environment (Fig. 6.6 shows a robot assembling itself from many modules that are inserted into a closed space and then carrying out a task). Many prototype robot systems have been built based on this framework (Fig. 6.7).

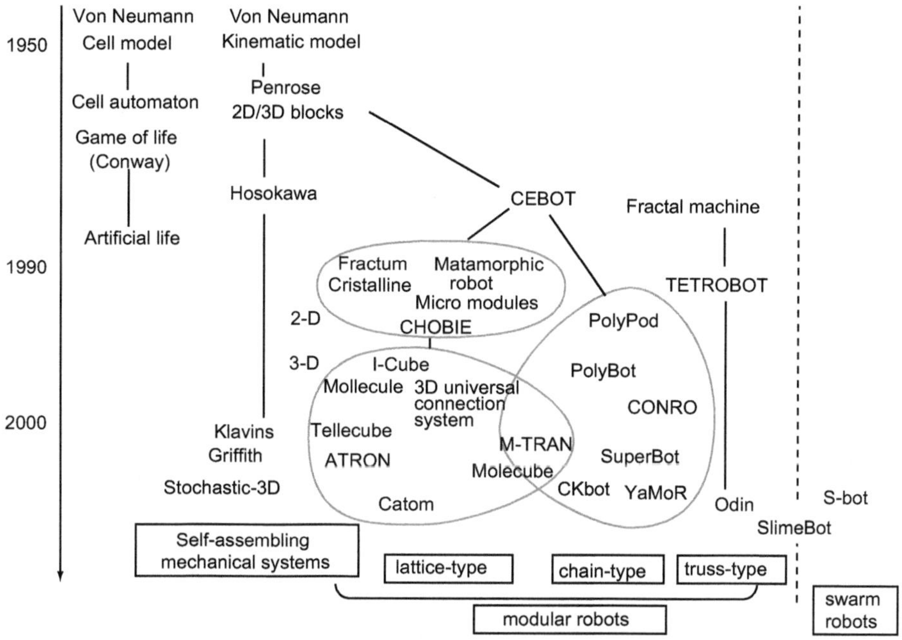

Fig. 6.5 History of modular robotics development

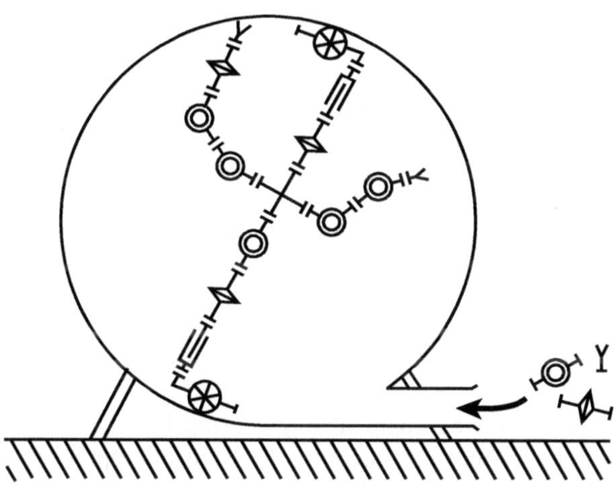

Fig. 6.6 Concept of CEBOT (Courtesy T.Fukuda, Nagoya U.)

Fig. 6.7 CEBOT series (Courtesy T.Fukuda, Nagoya U.)

The first version, called CEBOT Mark I, consists of two types of modules, those equipped with wheels for movement and those equipped with joints, grasping ability, and other functions. Various systems can be built from combinations of these modules [1]. CEBOT had many advanced aspects, such as introduction of connectors with large positioning error tolerance (Mark II) and docking experiments of hovering hexagonal modules (Mark III). In later years, the emphasis of his research has shifted to topics such as control of a swarm of mobile robots, applications of the CEBOT concepts to factory automation, and algorithms for evolution of cooperative behavior among the modules.

6.4.2 Truss-Type: Fractal Machine

Shigeru Kokaji of AIST's Mechanical Engineering Laboratory (MEL) developed a parallel computing system that included 64 8-bit microprocessors in 1980, and another more application-oriented parallel computing system MX2, whose processing unit consisted of 16 16-bit microprocessors, in 1986. His Fractal Machine is a two dimensional multiple-DOF mechanism controlled by MX2 [2]. This system is thoroughly based on the design principles of a distributed system in both its software and hardware.

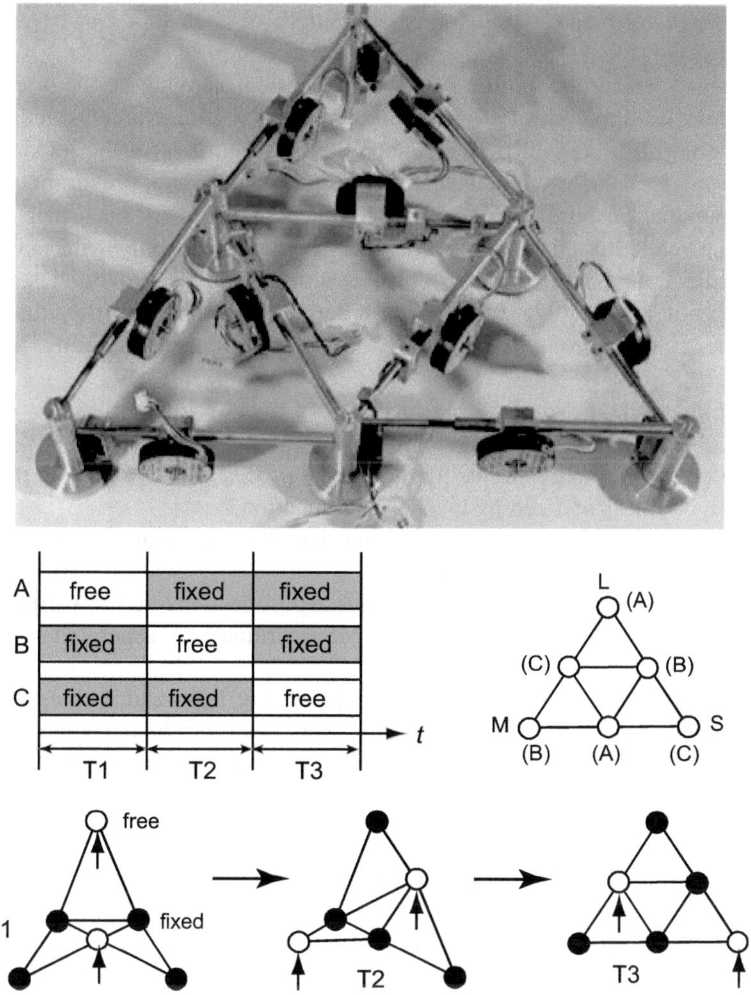

Fig. 6.8 Fractal Machine. Movement in one direction is realized by coordinated expanding/contracting of links and locking/unlocking of nodes in phases, depending on their positions in the LMS coordinates (Courtesy S.Kokaji)

Fractal Machine was built with hierarchy. The first order system of Fractal Machine consists of a triangular linkage mechanism consisting of three telescopic links, the second order system consists of three first order systems, and so on, so that an n-th order system can be constructed recursively (Fig. 6.8). The entire system is a telescopic link mechanism with multiple-DOF that has the fractal structure of the Sierpinski gasket. By employing a fractal truss structure whose links are connected by free rotating joints, the length of each link can be controlled independently.

In order to realize the motion of the whole system, moving in one direction for example, we need to control all the actuators in the system. Namely, stepping motors that control the length of links and solenoids that control the locking of the nodes (joints) to the ground have to be appropriately and synchronously controlled. For that purpose, one MX2 processor is assigned for each pair of a link and a joint (called a unit). These processors communicate only with neighboring units in the fractal structure. It is also assumed that, in the initial state, each unit does not know either the system order or its position within the system. Under these assumptions, control software with complete scalability[4] has been developed.

An outline of the Fractal Machine control is:

1. Wake-up: operation guaranteeing that all the processors have entered the operational mode. It is realized by local communication.
2. Synchronization: operation guaranteeing that all the units are synchronized to a single clock, with no phase difference.
3. Localization: by sending out commands from the units at the vertices of the structure, a coordinate system called LMS coordinates is generated on the fractal structure
4. Locomotion: time schedules of the contraction of links and locking/unlocking of nodes are determined according to their positions in the generated LMS coordinates, so as to realize desired motions such as movement in one direction and rotation in place (Fig. 6.8).

6.4.3 Truss-Type: TETROBOT

Arthur C. Sanderson at Rensselaer Polytechnic Institute developed TETROBOT, which has a multiple-DOF linkage structure like the Fractal Machine, but which is distinctive for its capability to form arbitrary three dimensional truss structures (Fig. 6.9) [3]. In a three dimensional truss, how to design the spherical joints where the ends of many links converge is the most important issue in the hardware design. They resolved this by introducing an off-centered pantograph mechanism. Control system of TETROBOT was basically a centralized one, though motion control by distributed processors was considered. This system is a Class 2 modular robot, which is not capable of changing its topological structure by itself, but which can be assembled manually into various configurations.

[4] Here, complete scalability means that a single control software program can control systems of any order, with constant communication traffic between units regardless of the order of the system.

Fig. 6.9 TETROBOT (Hamlin G, Sanderson A (1997) TETROBOT A Modular Approach to Reconfigurable Parallel Robotics, Springer ©1997 Springer)

6.4.4 Lattice-Type: Metamorphic Robot

In 1994, Gregory Chirikjian of Johns Hopkins University developed the Metamorphic Robot (Fig. 6.10) [4, 5]. Its module is a hexagonal two dimensional linkage mechanism with three servomotors for controlling shape transformation. Three more servomotors are used to drive the connecting mechanism. This is considered to be a system with hexagonal lattice-type modules. The mode of transportation between lattice points is rotation, but the problem of interference among modules is alleviated by the shape change of the module.

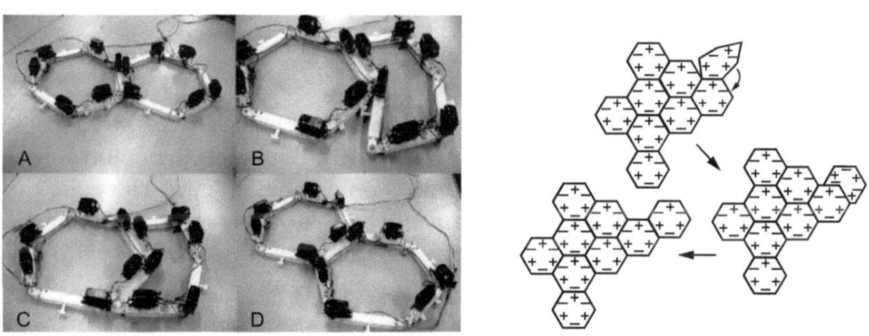

Fig. 6.10 Metamorphic robot. (Chirikjian GS (1994) Kinematics of a Metamorphic Robotic System, Proc IEEE Int Conf Robot Autom (ICRA), 1:449-455 ©1994 IEEE, Pamecha A, et al (1995) Design and Implementation of Metamorphic robot, Proc ASME Des Eng Tech Conf :18-22©1996 ASME)

6.4.5 Lattice-Type : Crystalline

The Crystalline robot [6, 7], developed by Daniela Rus of MIT, is composed of two dimensional lattice-type modules that expand and contract (Fig. 6.11). The lattice is square, and each module can expand to exactly double its standard length. Each module has actuators for horizontal and vertical expansion/contraction and four locking mechanisms, one for each face. The Crystalline prototypes were built using the rapid prototyping apparatus called FDM (Fused Deposition Modeling). Telescopic modules similar to those of Crystalline were also developed at the Xerox Palo Alto Research Center (TeleCube) and at Carnegie Mellon University.

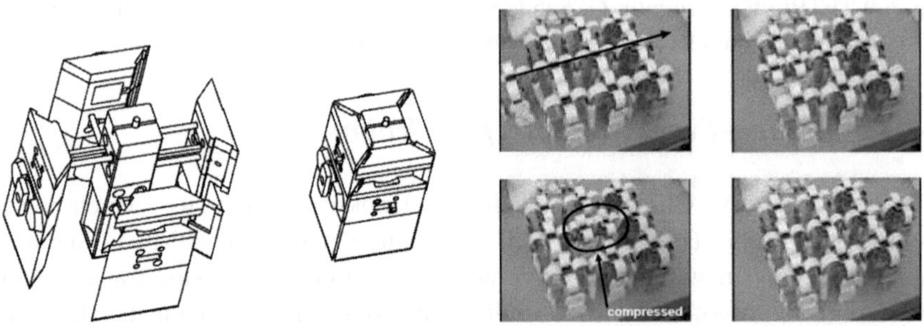

Fig. 6.11 Crystalline (Rus R, Vona M (1999) Self-reconfiguration Planning with Compressible Unit Modules, Proc. IEEE Int Conf Robot Autom (ICRA), 4:2513-2520 ©1999 IEEE, Butler Z, Rus D (2003) Distributed Planning and Control for Modular Robots with Unit-Compressible Modules, Int J Robot Res, 22(9):699-715 ©2003 IEEE)

6.4.6 Lattice-Type: Micro Modules

Eiichi Yoshida, a member of our group at MEL, developed Micro Modules, using actuators with shape memory alloy (SMA) [8-10]. Although the control circuitry is provided externally, the size of a module is only about 30 mm. These are the lattice-type modules based on a square lattice, which move by rotation. The rotational actuators and connecting pins are driven by heating SMA springs with electric current (Fig. 6.12).

6.4.7 Lattice-Type: CHOBIE

Norio Inou of Tokyo Institute of Technology developed a lattice-type modular robot CHOBIE [11]. He focused on a phenomenon of morphogenesis that he described as follows: "when an external force is applied to biological tissues providing mechanical support such as bones and muscles, the morphology and the

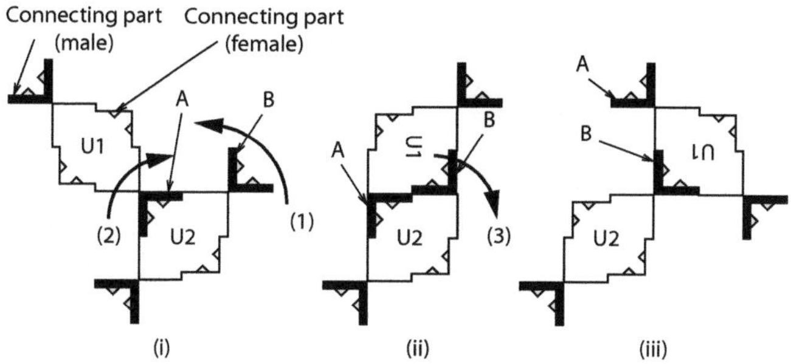

(a) Basic motion of two units

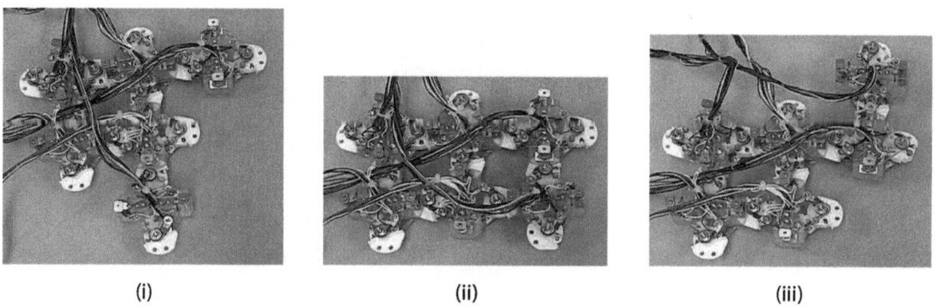

(b) Reconfiguration of many SMA units

Fig. 6.12 Micro Modules

strength distribution of those tissues changes adaptively according to the applied force so that excessive stress on the tissue is avoided." Accordingly, he proposed a method for determining the optimum structure to deal with given external forces using a structure generating algorithm which is a combination of finite element analysis and cellular automaton methods. The two dimensional module was developed for the purpose of giving building structures the capability to adapt to stress (Fig. 6.13). It is a system consisting of robotic modules connected by sliding connection mechanisms, and it is capable of reconfiguring its structure so that the stress detected by a strain gauge is averaged out though the entire system. Since its motion is simple sliding only, the sequencing of motions becomes rather complex.

6.4.8 Lattice-Type: Three Dimensional Universal Connection System

The system developed by the authors after Fractum was the Three Dimensional Universal Connection System [12]. The module of this system is a cube with a rotatable connector arm on each face, and the modules are arranged in a cubic lattice (Fig. 6.14).

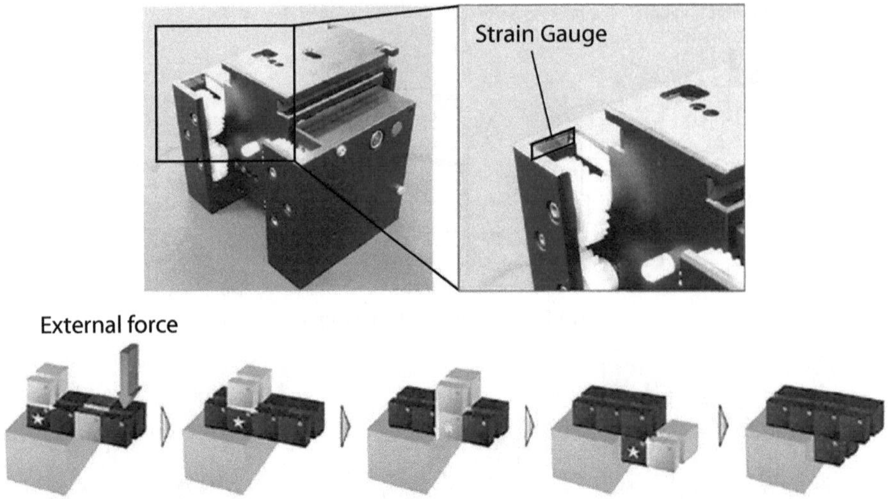

Fig. 6.13 Module CHOBIE (Suzuki Y, et al (2008) Reconfigurable Modular Robot Adaptively Transforming a Mechanical Structure, J Robot Soc Japan, 26(1):74-81 ©2008 RSJ)

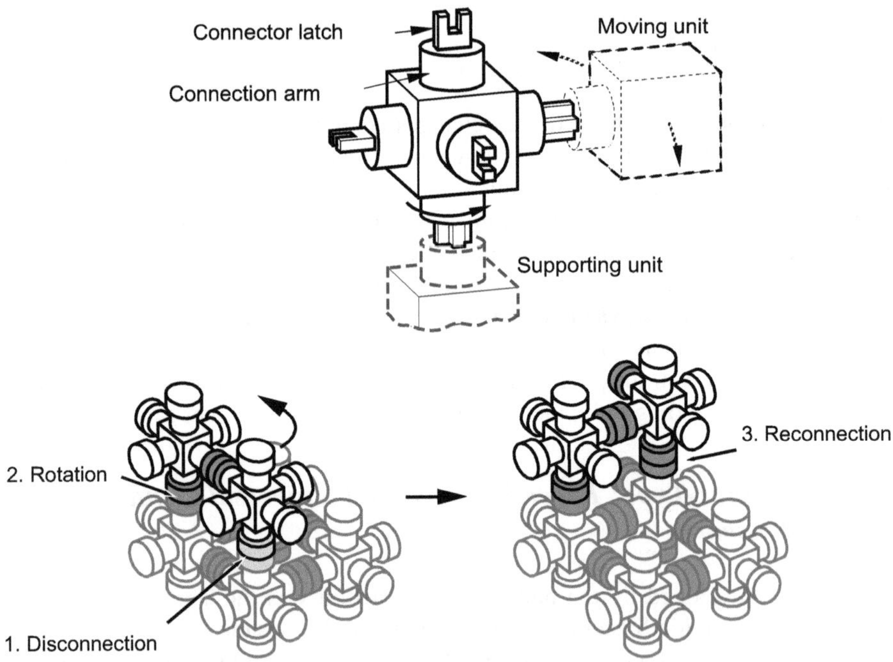

Fig. 6.14 Three Dimensional Universal Connection System

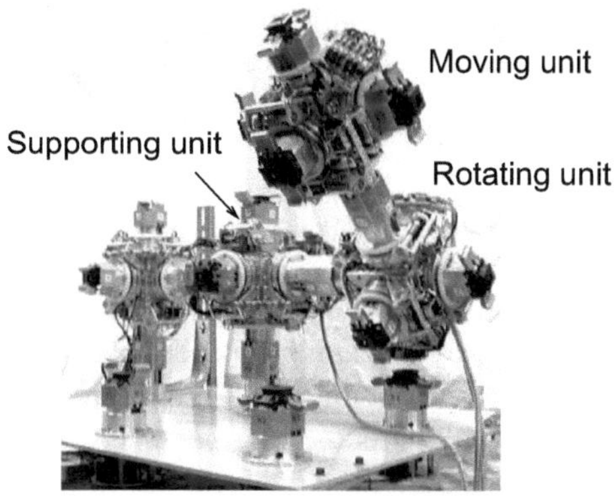

(a) Unit transfer motion

(b) Unit transfer completed

Fig. 6.15 Module lifting experiment using the Three Dimensional Universal Connection System

Connection and rotation of each arm can be independently controlled. Each module is given only one motor, but it also has six electromagnetic clutches and six solenoids that provide a mechanism for selecting its motion. The module design ended up being mechanically too complex, but one module can lift another module (Fig. 6.15). A distinctive feature of the module is its hermaphrodite connectors (Fig. 6.16). The six arms of a module are all identical, and therefore the algorithm for assembly can be simplified. It is possible to build simple target configurations according to Algorithm (I), described in the previous chapter, when this program is extended to three dimensions [13]. The connector is driven about a single axis, but it can tolerate positioning errors because its connecting claw easily catches the head of another connector, and once the joining operation is complete, the link becomes very firm[5].

6.4.9 Lattice-Type: Molecule

Daniela Rus at MIT developed a system called Molecule, whose module consists of two cubic units connected by a rotating link (double unit type) [14]. The modules come in two types, an active module with gripper mechanisms driven by motors, and a passive module with parts to be gripped (Fig. 6.17). Consequently, the mechanical complexity of the system is reduced, but in turn the algorithm for assembling it is more complex.

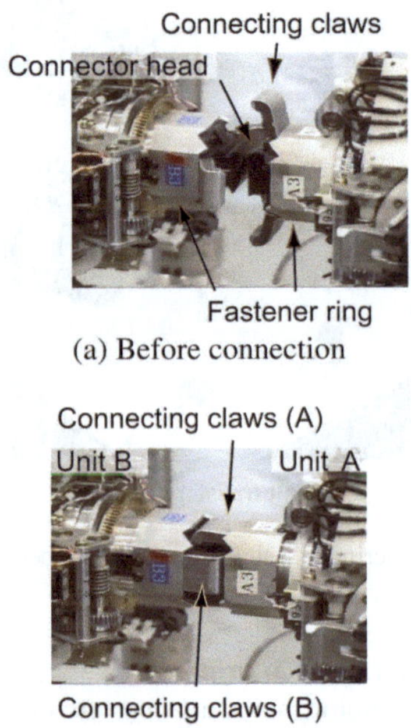

(a) Before connection

(b) In the process of connection

Fig. 6.16 Hermaphrodite connectors

[5] When a system consists of many modules, accumulation of errors is inevitable. Therefore, reconfiguration actually becomes more difficult when the modules are rigid and the connections are strong. In such a case, it may be necessary to relax the rigidity of the entire system to eliminate internal warping. See Section 9.3.1.

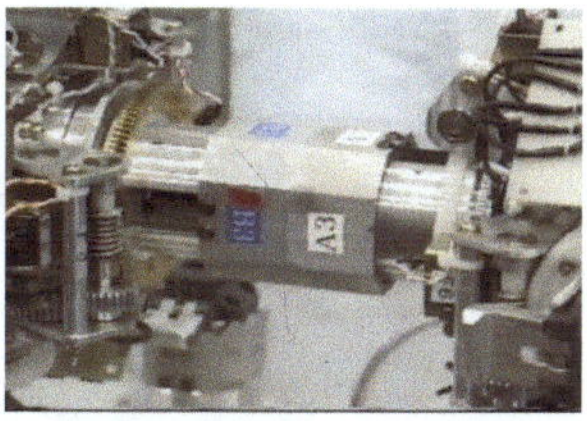

(c) Connection completed

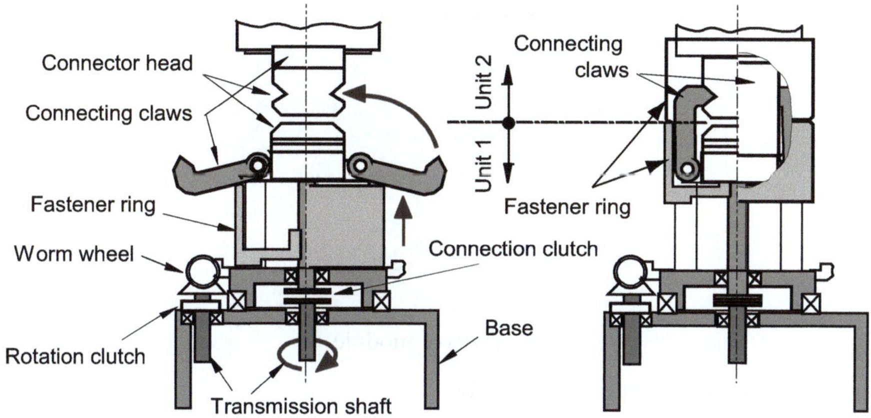

(d) Released (e) Connected

Fig. 6.16 *(continued)*

6.4.10 Lattice-Type: ATRON

The system ATRON, developed by Henrik Lund of the University of Southern Denmark, is a modular robot whose spherical module consists of two half-spheres (Fig. 6.19) [15]. The half-spheres can rotate relative to each other, and the module has connection mechanisms in eight directions by which it can connect with and disconnect from neighboring modules.

Fig. 6.20 shows a demonstration of seven ATRON modules traveling on a surface with an obstacle [16]. When the robot in a vehicle shape approaches an obstacle (the leftmost image), it changes shape and crawls over the obstacle using a method called cluster walk, in which the modules rotate to change the positions where they contact each other. An ATRON module is designed so as to have the minimum degree of freedom. On the other hand, due to this simplicity, the processes of motion generation and reconfiguration require coordination of many modules, and also there are many geometric constraints to be considered.

(a) Active module

(b) Passive module

Fig. 6.17 Molecule (Courtesy D. Rus, MIT)

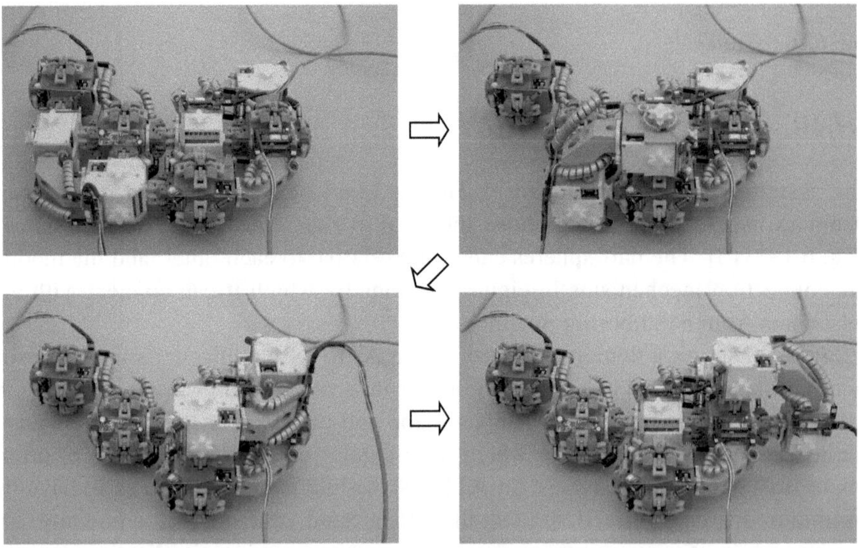

Fig. 6.18 Reconfiguration experiment of Molecule (Courtesy D. Rus, MIT)

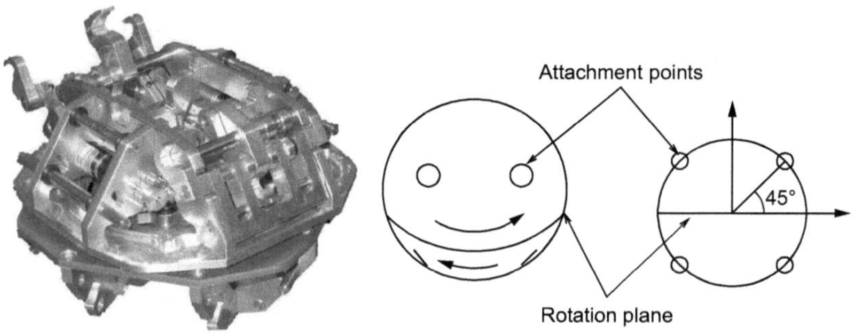

Fig. 6.19 Mechanical parts of the ATRON (Jorgensen MW et al. (2004) Modular ATRON: modules for a self-reconfigurable robot, Proc IEEE/RSJ Int Conf Intel Robot Syst (IROS 2004), 2:2068-2073 ©2004 IEEE)

Fig. 6.20 ATRON movement (cluster walk) (Østergaard EH, et al (2006) Design of the ATRON lattice-based self-reconfigurable robot, Auton Robot, 21(2):165-183 ©2006 Springer)

6.4.11 Lattice-Type: Molecube

The self-reproducing module, Molecube, developed by Hod Lipson at Cornell University is a cubic module split into two halves along the (1 1 1) plane, so that one half can rotate relative to the other half around an axis perpendicular to this cross section [17]. This plane has 3-fold rotational symmetry, so that every 120 degrees of rotation the module returns to be in the shape of a cube (Fig. 6.21(a)). The faces of the cubes can be connected by magnetic attraction. A vertical column formed by the modules can be changed by changing the rotation angle for each module (Fig. 6.21(b)). The sequence in Fig. 6.21(c) shows a process of self-reproduction by these modules. A four-module column (the completed configuration) connected to the base plate assembles the same configuration as itself, by picking up material modules from a specified feeding location one by one and piling them up at a specified location next to itself. This is a self-reproduction process of a Class 4 robot.

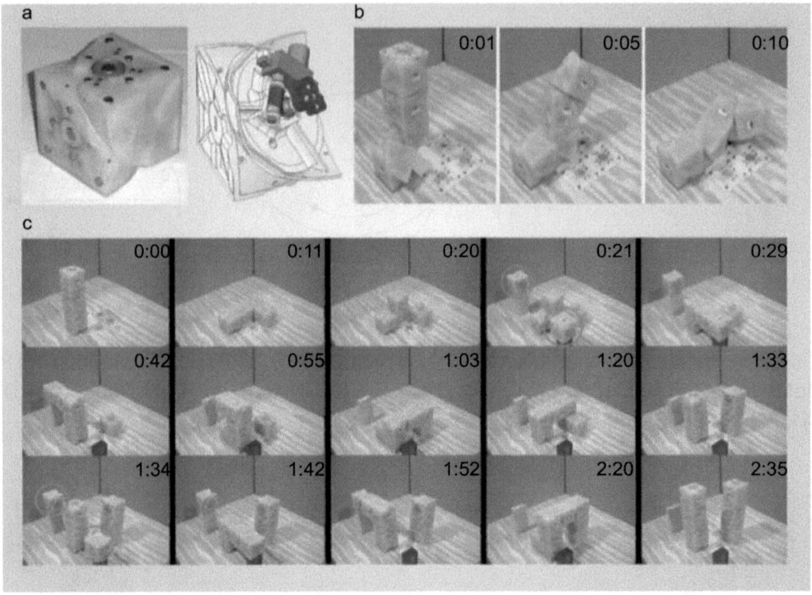

Fig. 6.21 Lipson's self-reproducing modular robot (Zykov V, Mytilinaios E, Adams B, Lipson H (2005) Self-reproducing machines, Nat, 435:163-164 ©2005 NPG)

6.4.12 *Chain-Type: PolyPod and PolyBot*

PolyPod (Fig. 6.22) is a chain-type modular robot developed by Mark Yim of the University of Pennsylvania, which consists of two types of modules: one capable of quasi-rotational motion by changing the shape of its link mechanisms, and the other for branching [18]. Since the modules have to be connected manually, PolyPod is a Class 2 modular robot. The main purpose of this robot is motion generation, and it can realize various periodic motions by controlling the joint angles of all modules in a synchronized cycle.

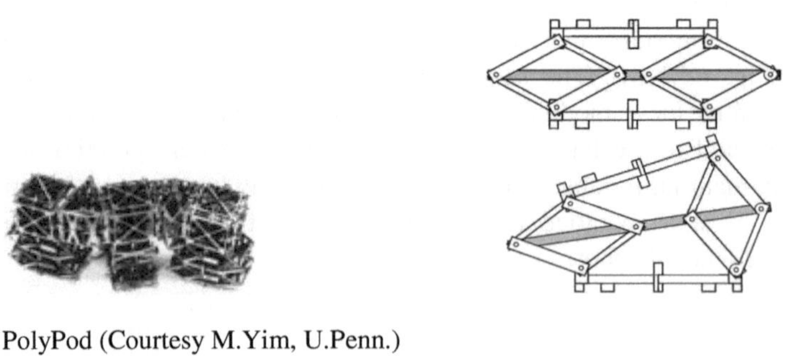

Fig. 6.22 PolyPod (Courtesy M.Yim, U.Penn.)

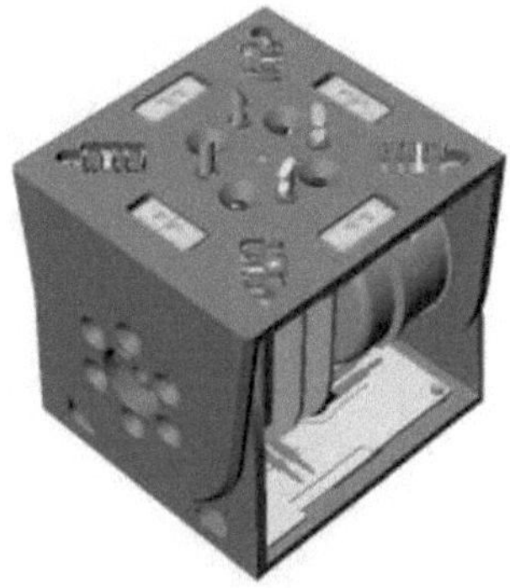

(a) A module with automatic connection mechanism

(b) A high torque module (manually screwed)

Fig. 6.23 PolyBot (Yim M, et al (2002) Connecting and disconnecting for chain self-reconfiguration with PolyBot, IEEE/ASME Trans mechatron, 7(4):442-451 ©2002 IEEE)

Another series with a similar name, PolyBot, was developed by the same group. Instead of the link mechanisms of PolyPod, it has a simple structure in which a motor is connected directly to the module's axis of rotation. Some models of PolyBot are equipped with automatic connection mechanisms and thus are Class 3 modular robots (Fig. 6.23(a)). (The module shown in Fig. 6.23(b) is a high torque type which requires connection by hand.) As we mentioned in Section 6.2, connecting chain-type modules to each other requires precise positioning, and for that purpose, PolyPod is capable of distance measurement, using multiple LEDs and phototransistors installed on the connecting surface [19]. However, reconfiguration requires precise alignment of connection parts, and is difficult to automate.

6.4.13 Chain-Type: CONRO and Superbot

The CONRO module, developed by Wei-Min Shen of the University of Southern California, is a chain-type module that has two degrees of freedom, pitch and yaw (Fig. 6.24) [20]. As for the connection topology, each module has three male connectors and one female connector, and an SMA actuator is embedded in the

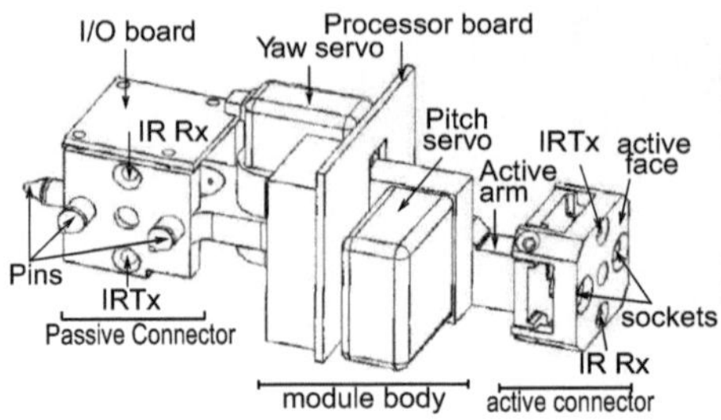

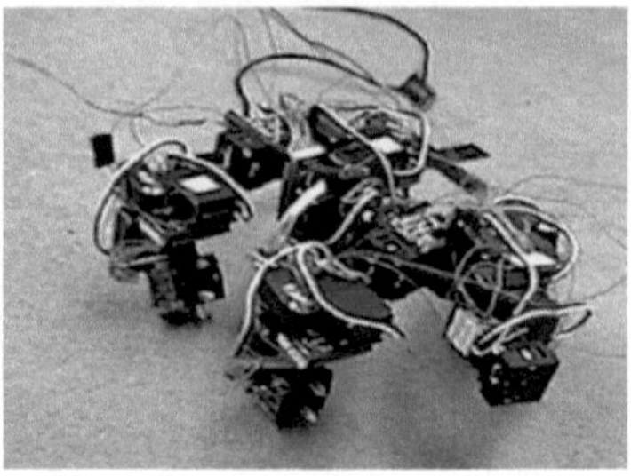

Fig. 6.24 CONRO (Castano A, Behar A, Will PM (2002) The Conro Modules for Reconfigurable Robots, IEEE/ASME Trans Mechatron, 7(4):403-409 ©2002 IEEE)

Fig. 6.25 SuperBot (Courtesy W.M. Shen, USC)

female connector, making automatic disconnection possible. However, like PolyBot, connection of modules is difficult without remote control by a human operator.

The same group developed the SuperBot module (Fig. 6.25) as an advanced version of CONRO. In addition to the two rotations of a CONRO module, it has a DOF for twisting at a link in the middle. Also, more faces of the module are used for connection, three each for male and female. The shape of this module is similar to M-TRAN which we discuss in the next chapter.

6.4.14 Lattice-Type: Catom

Claytronics is a project run by Seth Goldstein at Carnegie Mellon University and Intel aiming at developing a three dimensional tangible display device for industrial designing. It is categorized as a lattice-type modular robot, and in order to provide shapes to designers, it aims to achieve detailed shape formation and dynamic modification. The goal is to have millions of 300 μm modules moving independently in coordination, something like in "Terminator 2". At present though, a prototype of two dimensional module called Catom (Fig. 6.26), which changes its connection position using 12 electromagnets, is 4.4 cm in diameter [21].

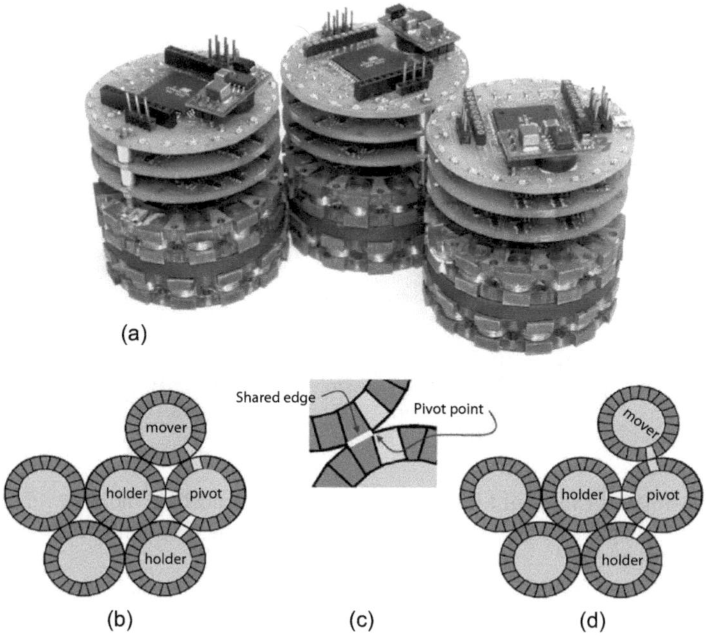

Fig. 6.26 Catom (Kirby BT, et al (2007) Modular Robots Using Magnetic Force Effectors, Proc IEEE Int Conf Intel Robot Syst (IROS) ©2007 IEEE)

6.4.15 Amorphous-Type: SlimeBot

The two dimensional modular robot SlimeBot is under development by Akio Ishiguro at Tohoku University [22]. The design of SlimeBot is inspired by slime molds. Circular modules have Velcro tapes placed around their side wall so that modules become connected at any point on their perimeter where they come in contact. They become disconnected when a certain pulling force is applied. The modules are also equipped with expanding/contracting actuators so that the shape of module can be deformed. The processor in each module has a van der Pol oscillator to control the oscillatory deformation of the modules. By using the entrainment phenomenon (see Section 8.3) among oscillators and symmetry breaking by external light input, a collection of many modules generates amoeboid motions, realizing locomotion. This collective locomotion is able to avoid obstacles as real

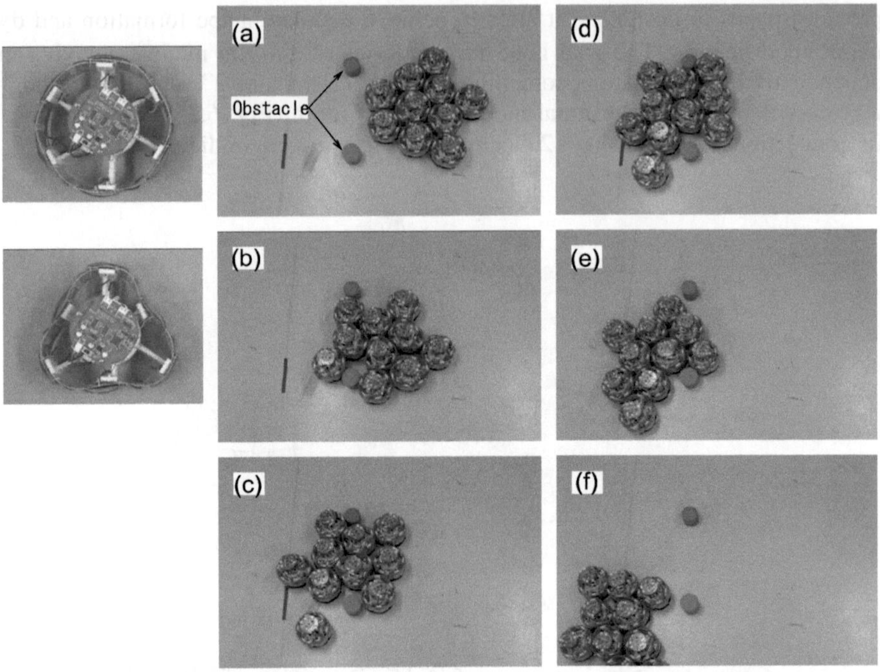

Fig. 6. 27 SlimeBot (Shimizu M, Ishiguro A (2009) An amoeboid modular robot that exhibits real-time adaptive reconfiguration, Proc IEEE/RSJ Int Conf Intel Robot Syst, 1496-1501 ©2009 IEEE)

slime molds do. SlimeBot has unique properties not found in other modular robots, in particular physical and mechanical interactions between modules that make essential contributions to the formation of order in the whole system.

6.5 Hybrid Type Combining Lattice and Chain

Among all the prototypes we looked at in this section, the lattice type modular robots and the amorphous-type one which we discussed last are the examples that can be considered to be self-organizing systems of Classes 3 and 4. In the research of modular robots of truss-type and chain-type, the emphasis is not so much on emergence of functions by self-organization but more on distributed control architectures and motion control by multi-DOF mechanisms.

With many of the lattice-type prototypes, however, only basic experiments using a small number of modules have been done, because of the complexity of the modules. For instance, with the Three Dimensional Universal Connection System we developed, we only did reconfiguration experiments using four modules, and thus had to make simulations to confirm self-organization of large scale systems.

This led us to a search for a more realistic design, a hybrid type module with characteristics of both the lattice-type and chain-type.

In the next chapter, we first give details of self-reconfigurability, a function realized by M-TRAN due to its lattice-type aspects. In Chapter 8, we focus on the motion control of M-TRAN as a chain-type modular robot, and discuss motion acquisition based on the self-organization principle.

References

[1] Fukuda, T., Nakagawa, S.: A Study on Dynamically Reconfigurable Robotic Systems. Trans. Japan. Soc. Mech. Eng. C 55(509), 114–118 (1989) (in Japanese)

[2] Kokaji, S.: A Mechanism of Very Many Degrees of Freedom and a Distributed Control System. J. of Japan Soc. Precis. Eng. 54(10), 1921–1926 (1988) (in Japanese)

[3] Hamlin, G., Sanderson, A.: TETROBOT A Modular Approach to Reconfigurable Parallel Robotics. Springer (1997)

[4] Chirikjian, G.S.: Kinematics of a Metamorphic Robotic System. In: Proc. IEEE Int. Conf. Robot. Autom. (ICRA), vol. 1, pp. 449–455 (1994)

[5] Pamecha, A., et al.: Design and Implementation of Metamorphic robot. In: Proc. ASME Des. Eng. Tech. Conf., pp. 18–22 (1995)

[6] Rus, R., Vona, M.: Self-reconfiguration Planning with Compressible Unit Modules. In: Proc. IEEE Int. Conf. Robot. Autom. (ICRA), vol. 4, pp. 2513–2520 (1999)

[7] Butler, Z., Rus, D.: Distributed Planning and Control for Modular Robots with Unit-Compressible Modules. Int. J. Robot. Res. 22(9), 699–715 (2003)

[8] Yoshida, E., et al.: Miniaturization of Self-Reconfigurable Robotics System using Shape Memory Alloy Actuator. J. Robot. Mechatron 12(2), 96–102 (2000)

[9] Yoshida, E., et al.: Micro Self-Reconfigurable Modular Robot Using Shape Memory Alloy. J. Robot. Mechatron 13(2), 212–219 (2001)

[10] Yoshida, E., et al.: Miniaturization of Self-Reconfigurable Robotic System using Shape Memory Alloy Actuator. J. Robot. Mechatron 12(2), 96–102 (2000)

[11] Suzuki, Y., et al.: Reconfigurable Modular Robot Adaptively Transforming a Mechanical Structure. J. Robot. Soc. Japan 26(1), 74–81 (2008)

[12] Kurokawa, H., et al.: A Three-Dimensional Self-Reconfigurable System. Adv. Robot. 13(6), 591–602 (2000)

[13] Yoshida, E., et al.: Self-Assembly and Self-Repair of 3-D Structure by an Autonomous Distributed Machine. Trans. Soc. Instrum. Control Eng. 35(11), 1421–1430 (1999) (in Japanese)

[14] Kotay, K., et al.: The self-reconfiguring robotic molecule. In: Proc. IEEE Int. Conf. Robot. Autom., vol. 1, pp. 424–431 (1998)

[15] Jorgensen, M.W., Ostergaard, E.H., Lund, H.H.: Modular ATRON: modules for a self-reconfigurable robot. In: Proc. IEEE/RSJ Int. Conf. Intel. Robot. Syst. (IROS 2004), vol. 2, pp. 2068–2073 (2004)

[16] Østergaard, E.H., et al.: Design of the ATRON lattice-based self-reconfigurable robot. Auton. Robot. 21(2), 165–183 (2006)

[17] Zykov, V., Mytilinaios, E., Adams, B., Lipson, H.: Self-reproducing machines. Nat. 435, 163–164 (2005)

[18] Yim, M.: Locomotion with a Unit Modular Reconfigurable Robot., Ph.D. Thesis, Dept. Mech. Eng. Stanford Univ. (1994)
[19] Yim, M., et al.: Connecting and disconnecting for chain self-reconfiguration with PolyBot. IEEE/ASME Trans. Mechatron 7(4), 442–451 (2002)
[20] Castano, A., Behar, A., Will, P.M.: The Conro Modules for Reconfigurable Robots. IEEE/ASME Trans. Mechatron 7(4), 403–409 (2002)
[21] Kirby, B.T., et al.: Modular Robots Using Magnetic Force Effectors. In: Proc. IEEE Int. Conf. Intel. Robot. Syst (IROS), pp. 2787–2793 (2007)
[22] Shimizu, M., Ishiguro, A.: An amoeboid modular robot that exhibits real-time adaptive reconfiguration. In: Proc. IEEE/RSJ Int. Conf. Intel. Robot. Syst., pp. 1496–1501 (2009)

Chapter 7
Robotic Metamorphosis

Abstract. M-TRAN is a three dimensional modular robotic system, having features of lattice-type and chain-type systems; it can self-reconfigure as a lattice-type modular robot, and can make versatile robotic motions as a chain-type modular robot. In this chapter, we discuss its lattice-type features: design principles, metamorphosis by small numbers of modules, and distributed self-reconfiguration by large numbers of modules. Throughout this section, the term "metamorphosis" is used interchangeably with the term "self-reconfiguration".

7.1 System Design

M-TRAN (Modular TRANsformer) is a three dimensional lattice-type modular robot [1, 2]. Many of the lattice-type modular robots discussed in the previous chapter have very complicated designs because they have many degrees of freedom, so actually building them was hard, and only small scale experiments have been made. For instance, a module of our Three Dimensional Universal Connection System (Fig. 6.14) has all the symmetry of a cube, and is equipped with six rotating arms each with connection capability, summing up to 12 degrees of freedom [3]. As a consequence, the prototype module we built was very large and heavy, mechanical connections between modules had to be made sturdy, and complex mechanisms were required to enable them to support each other.

On the contrary, a module of M-TRAN, consisting of two cube-like blocks joined together, has only two rotational degrees of freedom and only three of the six faces are capable of active connection. In this section, we explain the shape and basic functions of M-TRAN modules.

7.1.1 M-TRAN Module

7.1.1.1 Shape and Function

An M-TRAN module consists of two blocks, each of which has the shape of a half of a cube and a half of a cylinder joined together, and a link connecting these

S. Murata and H. Kurokawa: Self-Organizing Robots, STAR 77, pp. 131–171.
springerlink.com

blocks (Fig. 7.1). Two blocks and one link are joined using two parallel axes, each axis going through the center of each block. Each block is allowed 180 degree rotation about its axis with regard to the link. Hereafter, these axes are called *joints,* following robotics terminology.

The three flat faces of each block actively make connection with faces of other modules. There are two shapes of these faces, one square and the other not, but their connection function is the same, meaning that there are four ways for two faces to connect so that their circumscribed squares coincide.

The M-TRAN system is a homogeneous modular robot, i.e. consisting of identical modules. Various configurations can be formed using these modules, exploiting the six connection faces of each module and the freedom in the direction of connections. Once modules are connected in a particular configuration, the whole is capable of robot-like motions by maneuvering its many joints. For example, Fig. 7.2(a) is a linear configuration where modules are connected in series with all the joint axes aligned in parallel. If we generate a propagating wave by moving the joints in a coordinated manner, the whole line can move forward on a surface. Similarly, (b) is a configuration for rolling, (c) for a manipulator, and (d) for a quadruped walking robot.

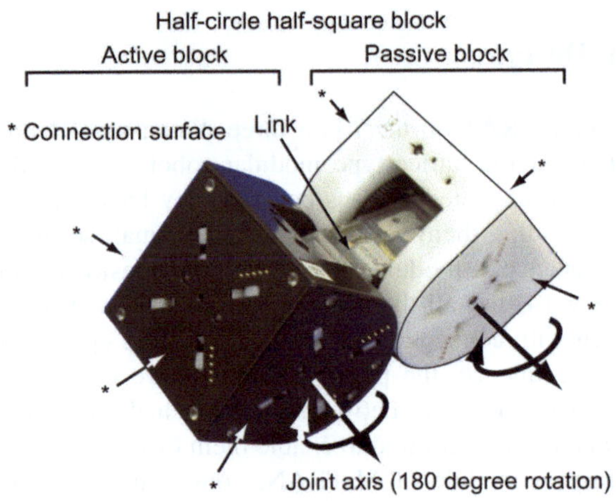

Fig. 7.1 M-TRAN module

The half-cube half-cylinder block can be inscribed in a cube, and this shape is obtained by shaving two connected cubic blocks until each can rotate without interfering with the other. Connected modules can fit in a cubic lattice, each block put in a unit cube of the lattice with each face in contact with the adjacent module's face, as shown in Fig. 7.3. We call this an *in-lattice state*. The joint angle is considered to be 0° when the semi-cylinders of two blocks are facing each other like the white blocks A and B in Fig. 7.3, and each block can move ±90°. This

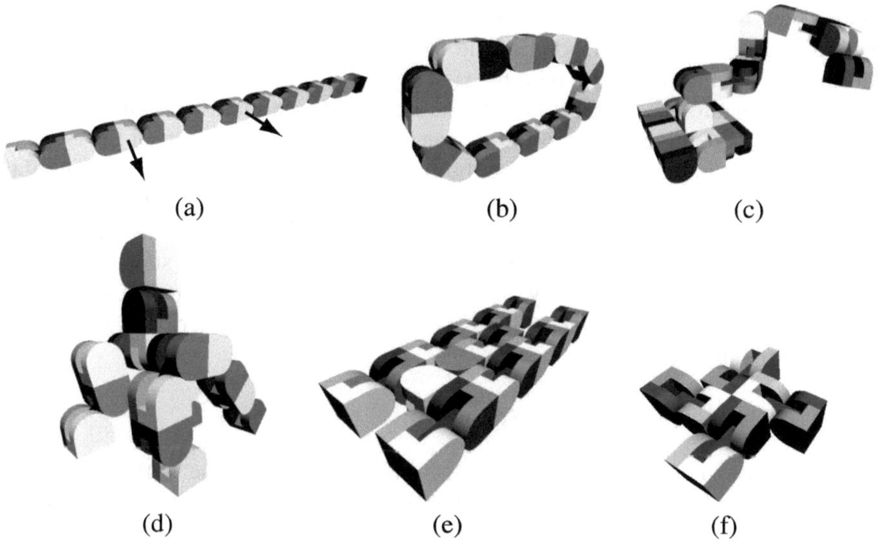

Fig. 7.2 Examples of configurations of M-TRAN modules

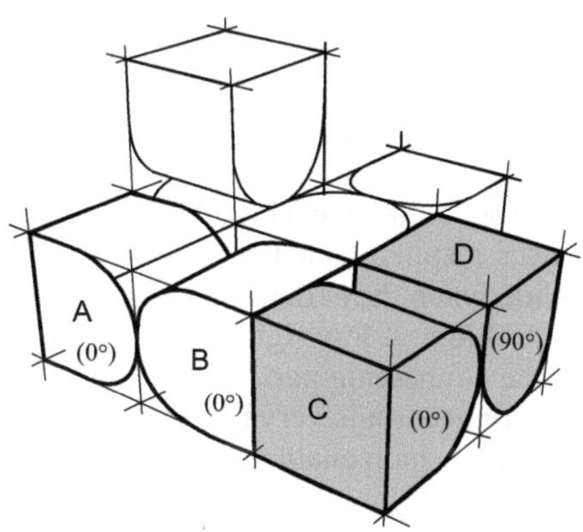

Fig. 7.3 In-lattice state

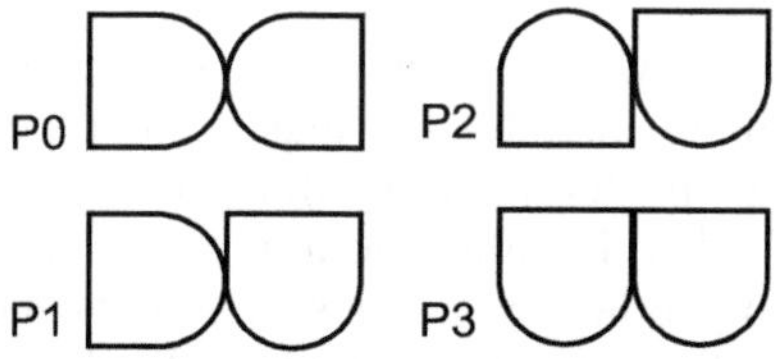

Fig. 7.4 Base states

means that an in-lattice state consists of modules whose joint angle is 0 or ±90°, and each module is in one of the four *base states* shown in Fig. 7.4, disregarding the orientation of the module as a whole.

Each connection face of M-TRAN can change or maintain its connections. M-TRAN modules change their configuration by disconnecting a part of the configuration, and moving and reconnecting the freed modules and blocks. As we explain below, each action of reconfiguration is basically carried out in an in-lattice state. The configuration of the quadruped robot in Fig. 7.2(d) turns into an in-lattice state in Fig. 7.2(e) when the legs are flattened out. In an in-lattice state, a module's state is discrete, characterized by its base state and the connection status of its faces. In principle, the metamorphosis procedure can be described by a finite sequence of state transitions.

7.1.1.2 Characteristics of the Shape

There are two major characteristics of M-TRAN modules; one is that the blocks connect at their faces, and the other is that the two axes of their link are in parallel. We discuss the first here, and explain the latter in Section 7.2.2.

First of all, each block of an M-TRAN module is designed to be as close to a cube as possible so that it has the maximum volume possible. This retains not only more room for installing machinery and circuitry, but also rigidity of the module body. Next, the two blocks of each module are always in contact at their curved surfaces no matter what their angles, making them resistant against pressure and distortion. Moreover, in in-lattice states, neighboring modules are either connected or merely in contact, which in either case prevents modules from being crushed or deformed when stacked up.

In the case of modular robots like the Three Dimensional Universal Connection System, a module occupies a sphere, smaller than the cubic unit of the lattice, so that modules do not collide when they rotate. As a result, the connector parts have to be small, causing the problem of low rigidity. In the case of M-TRAN, since an entire face of a block is the connecting part, and this face has the maximum possible surface area, thus the connection is very rigid. Moreover, there is no protrusion in the connection mechanism, which enables various motions.

7.1.2 Basic Motions

7.1.2.1 Motions on the Ideal Plane

Consider an ideal situation where a module is placed on a plane consisting of a layer of modules. A single module can make two types of motion in such a situation depending on its posture: rolling along a line (Fig. 7.5(a)) and pivoting over the surface (Fig. 7.5(b)). In either motion, one of the module's block moves from one lattice point to another by a series of actions such as releasing connection, controlling joint angles so as to move one of the blocks, and finally connecting to a new face.

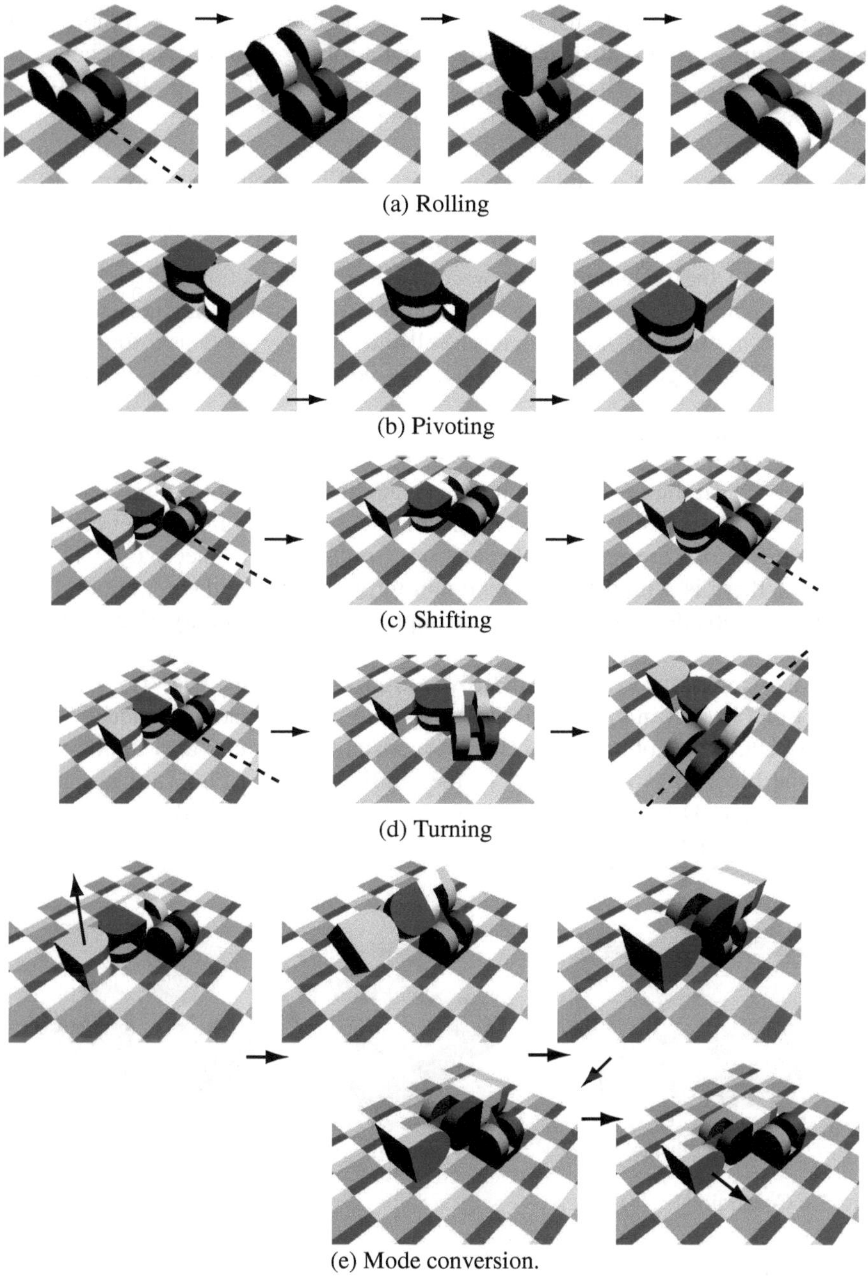

Fig. 7.5 Basic motions.

By the above two motions, the direction of module's joint axes do not change. Also, the line of rolling does not change, and a rolling module cannot change to a pivoting module by itself. These limitations are resolved by coordination of two modules. In Figs. 7.5(c) and (d), by pivoting motion of one of the modules, the line of rolling shifts or turns at a right angle. By the coordinated motion in Fig. 7.5(e), the posture of a module is changed and rolling module is converted to a pivoting module. Note that, in order to do mode conversion under gravity, a module needs to be at least powerful enough to lift another module.

In most of the metamorphoses we discuss in later sections, there is no ideal plane available or it is sometimes necessary that two or more modules be lifted, so these basic motions are not sufficient. Still though, they serve as a foundation for metamorphosis, together with the collision avoidance motions we discuss next.

7.1.2.2 Constraints

There are constraints in module motions; modules must not collide with others during rotation (collision avoidance), the entire configuration must not come apart when connections are released (separation avoidance or connectivity constraint), and since there is a limit to the maximum torque of the hardware, only a limited number of modules can be lifted or supported under gravity (hardware constraints).

Let us explain the collision and division problems using Fig. 7.6. Obviously, the motion (1) of A in Fig. 7.6(a) is not possible because another module is blocking the path. The motion (2) of changing posture in place is also impossible because the block B is in the way. In the configuration in Fig. 7.6(b), if the two modules C and D move independently, the whole may be divided in two, so this should be avoided.

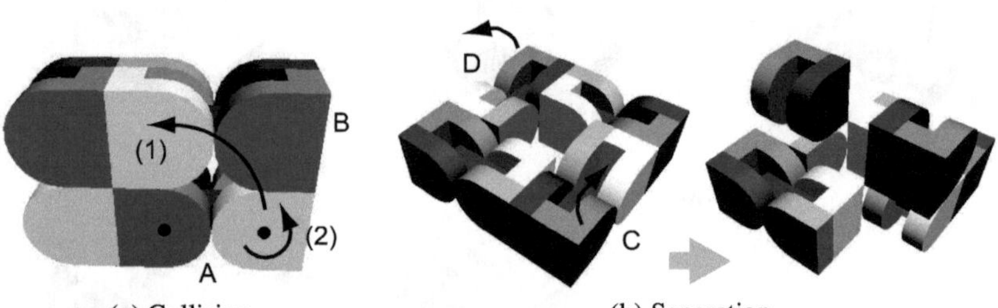

(a) Collision (b) Separation

Fig. 7.6 Constraints

The cube-like shape of the module's block makes the problem of collision likely, and designing the module to have parallel joints was meant to alleviate this problem. The module A in Fig. 7.7(a) collides with the adjacent module B if only one of its joints is driven. However, when the two parallel joints are driven at the

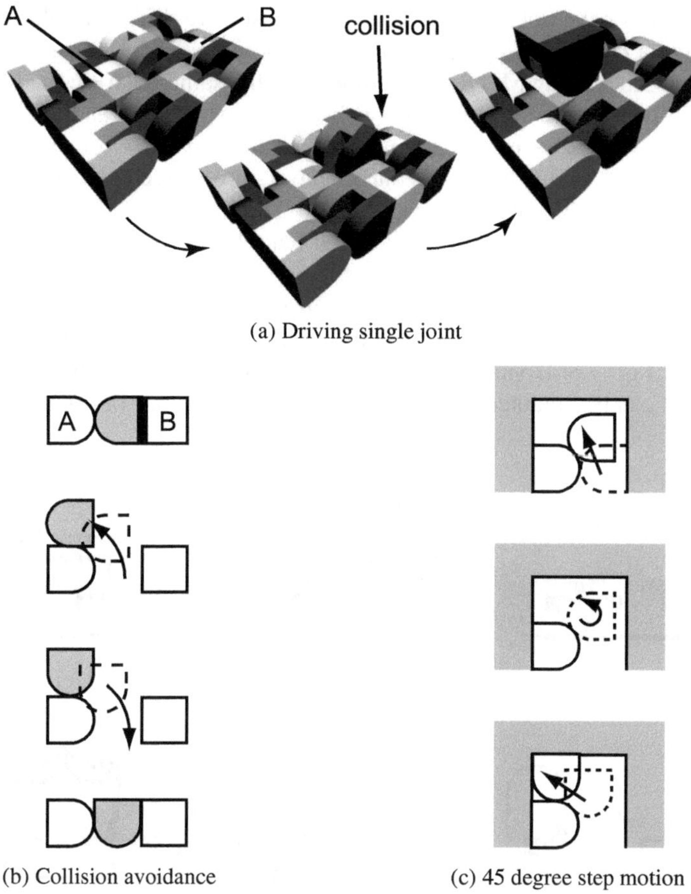

Fig. 7.7 Collision avoidance motions.

same time in the opposite directions, collision is avoided as in Fig. 7.7(b). Even in the case where a module is surrounded by other modules as Fig. 7.7(c), it is possible to move a block without collision by changing two joint angles simultaneously in 45° increments.

The number of connected modules that can be moved at a time depends on whether the motion is lifting against gravity or moving horizontally, and also depends on the shape of the modules to be carried. In reality, there is a limit to the joint torque, and moreover, structural deformation and limits on the precision of positioning also affect the success of connection.

7.1.3 Polarity

Although the two blocks of a module have exactly the same geometrical shape, they have different polarities, which makes the hardware simpler. This means that connection faces have polarity, and two faces must have different polarity to connect, just like a key and a keyhole, or north and south poles of a magnet. All three faces of each block of an M-TRAN module have the same polarity; the block with faces that actively make connections is called *active* (or male), and the block with faces that allow themselves to be connected is called *passive* (or female). In the figures and photos in this book, the darker blocks are active ones, and those with lighter color are passive ones.

As long as connection changes are carried out while strictly maintaining the in-lattice state, the polarity does not give rise to any problem. Just as a checker board is painted in two colors, M-TRAN can be arranged in a three dimensional cubic lattice so that any adjacent blocks have different colors. Note that the sequence of motion from Fig. 7.8(a) via (b) to (c) cannot happen in an in-lattice state, but a similar result can be obtained by the sequence in Fig. 7.8(a), (d), (e) and (f).

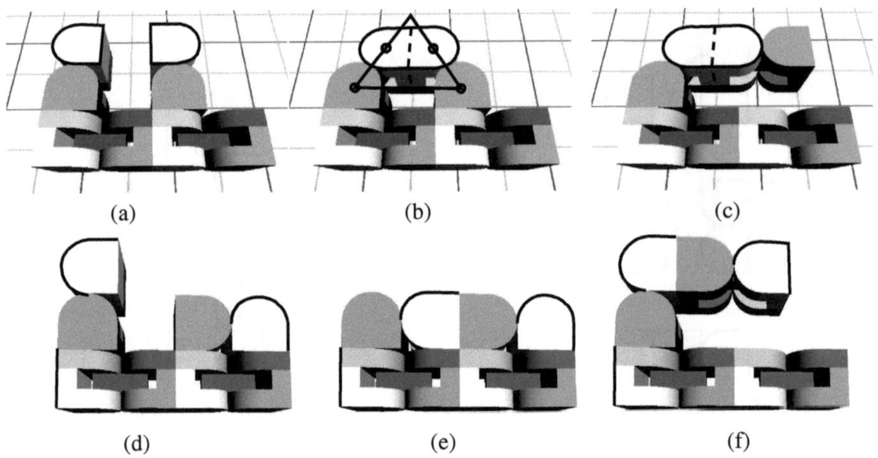

Fig. 7.8 Polarity problem. Since M-TRAN module connections can be made only between a white block and a grey block, the sequence from (a) to (c) via (b) is not allowed. However, a similar result (f) is obtained from (a), via (d) and (e).

7.1.4 Universal Assembly and Self-reconfiguration

Now assume that there are a sufficient number of modules lying on the plane that we had in Fig. 7.5. Each module can move to any position on the plane with any orientation by the basic motions in Fig. 7.5. By repeating this process according to an assembly plan, it is possible to build almost any two dimensional configuration. Procedures for building planar structures like those in Fig. 7.2(a), (e), and (f) can be easily created, as shown in Fig. 7.9. Moreover, if we use a three dimensional arm made of modules, three dimensional structures can also be built. By detaching each completed structure, various structures can be made one after another, as long as there are modules remaining. This is indeed a universal assembly machine.

However, in this case, there is a clear distinction between what is in charge of assembly and what is being assembled, which is not a very interesting kind of self-assembly. In general, the term *self-reconfiguration* implies the cooperative action of organically related modules to change their configuration without any external help. In the case of the above universal assembly, the assembly procedure for a target configuration is straightforward, whereas a procedure for self-reconfiguration, in which all the modules in the system take part in assembling the structure (we also call this *metamorphosis* hereafter), may not be simple.

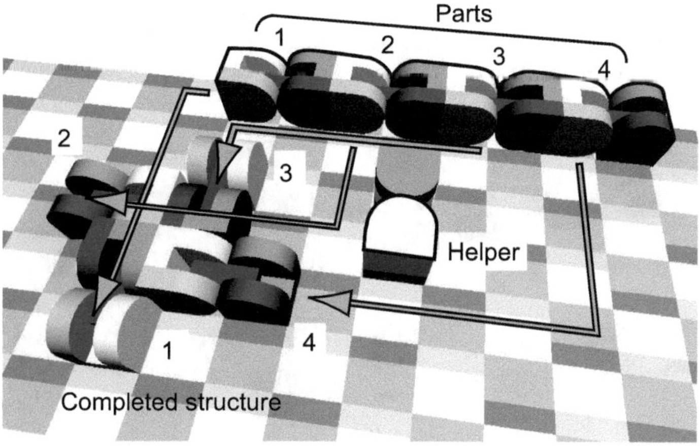

Fig. 7.9 Assembly procedure

7.2 Planning Metamorphosis Procedure

M-TRAN is a system each of whose modules functions as an autonomous subsystem controlling its own joint angles and connections. What we aim now is to make the connected modules interact so that they form three dimensional configurations in a self-organizing manner. Is it possible for the M-TRAN modules to transform into a target configuration like Fractum in Chapter 5?

The M-TRAN design sacrifices some of the degrees of freedom and symmetry of an ideal lattice type system. The resulting loss of function is to be compensated for by coordination between multiple modules, but it is not a simple task to determine how modules should coordinate their behavior. We have to face the difficulties presented by the geometry of the motion before considering distributed control.

In this section, we set aside the issues related to distributed control, and concentrate on the geometry of reconfigurations of M-TRAN modules.

7.2.1 *Search for Metamorphosis Procedures*

7.2.1.1 Reconfigurability

Consider metamorphosis between given two configurations consisting of the same number of modules, for example, four M-TRAN modules as in Fig. 7.10. Among many possible configurations, it is obvious that metamorphosis is impossible between the two configurations (a) and (c) in Fig. 7.10. This is because all the rotation axes are in parallel in (a), so that the modules can transform itself to a shape like (b), but cannot change the direction of their axes. On the other hand, configuration (d) has one module with a different axis direction. It is easy to see that (d) can transform to (e), and in fact, it can transform to (c) eventually. Similarly in the case of ten modules, metamorphosis from a configuration as in Fig. 7.2(a) to one as in Fig. 7.2(e) is not possible, but (e) can be reached if one module in (a) is twisted (see Section 7.2.2).

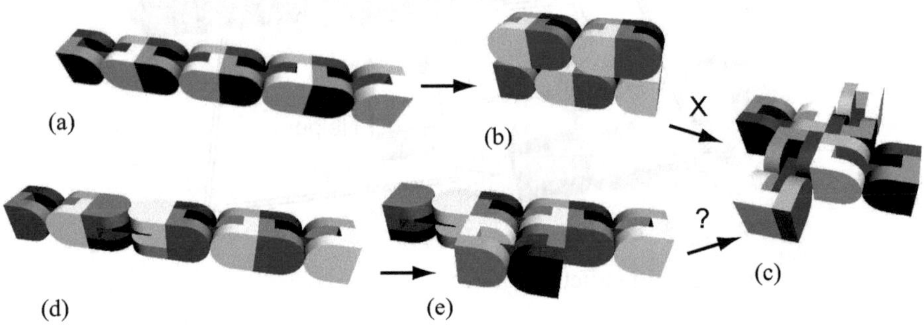

Fig. 7.10 Reconfigurability

7.2.1.2 Exhaustive Search

Let us next consider the search for a metamorphosis procedure. First, we need a method of representing the configurations and metamorphosis procedures [4-6]. There may be several methods of representation, and any of them can be used as long as two configurations can thereby be determined to be identical or not. One method is to align modules in the lattice space and use discrete coordinate values defining the position and orientation of each module. Another method considers only the connection topology, ignoring joint angles. Each module has six connection faces, and the relative orientation of the two connecting faces of each connection between modules is represented by a discrete number. Hence, the whole configuration can be represented by a graph. A metamorphosis procedure is a sequence of primitive actions: discrete changes in joint angles and connections.

The search process starts from the initial configuration, and primitive actions are carried out until the target configuration is reached. Each time there are a finite number of possible actions, by which possible configurations grow in number in the form of a tree structure whose root is the initial configuration [6]. It should be noted that the same configuration may be arrived more than once in the search process, and therefore it should be checked every time to see if the resulting configuration already appears in the tree or not, in order to avoid searching a branch more than once. It is also necessary to avoid collisions and check if the whole configuration is still connected.

Although there are many problem solving methods employing traversal of a tree or a graph structure to reach a target, the search space expands explosively as the number of modules increases. The complexity of this problem is considered to be NP-hard, so that searching cost increases exponentially with the number of modules [7].

The reachability problem, determining whether there is a metamorphosis procedure between given two configurations, can also be reduced to the same search problem. Moreover, even if a path is obtained by searching, whether it is optimal or not can only be determined by a full search[1].

7.2.1.3 Heuristics

Generally, searching for a metamorphosis procedure between arbitrarily chosen configurations often is in vain, because reachablility is not assured before searching. Therefore, we adopted a *heuristic* method to find various metamorphoses, aiming for a roughly defined goal of transformation between different types of configurations such as a legged one and a linear one. In this process, initial and target configurations are modified by trial-and-error while searching.

7.2.2 Metamorphosis between Mobile Robot Configurations

There are various configurations of M-TRAN modules which are mobile, and we searched for metamorphoses between them. Reconfigurations which we designed and verified by experiments are shown in Fig. 7.11.

7.2.2.1 Parallel Quadruped Form

Quadruped robots in Fig. 7.12(a) to (f) have legs each consisting of two modules with axes all in parallel. We call these parallel quadruped forms. We searched for a metamorphosis to a linear configuration from the configurations (a) - (f). It is obvious that (a) and (b) cannot reach a linear structure because all the joint axes are in parallel. But from (d), (e), or (f), reconfiguration to (g) is possible. An actual metamorphosis procedure from (e) to (g) was designed taking hardware constraints into consideration, and verified by experiments as shown in Fig. 7.13 (a).

---

[1] Distance measure between two configurations can be defined as the minimum number of actions needed to make this transformation. It cannot be obtained without full search and is not related to any geometrical similarity between two configurations. See [1, 8, 9].

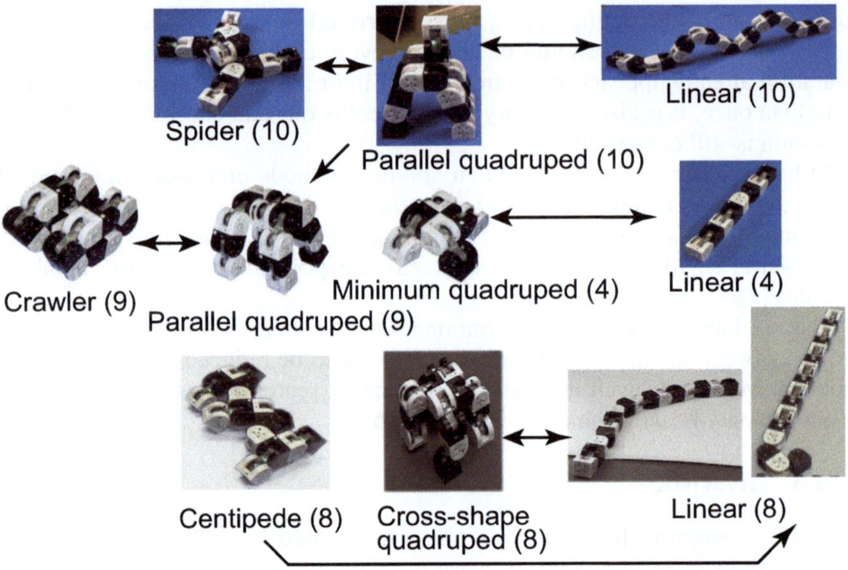

Fig. 7.11 Transitions between robot configurations. Below each configuration is its name and number of modules.

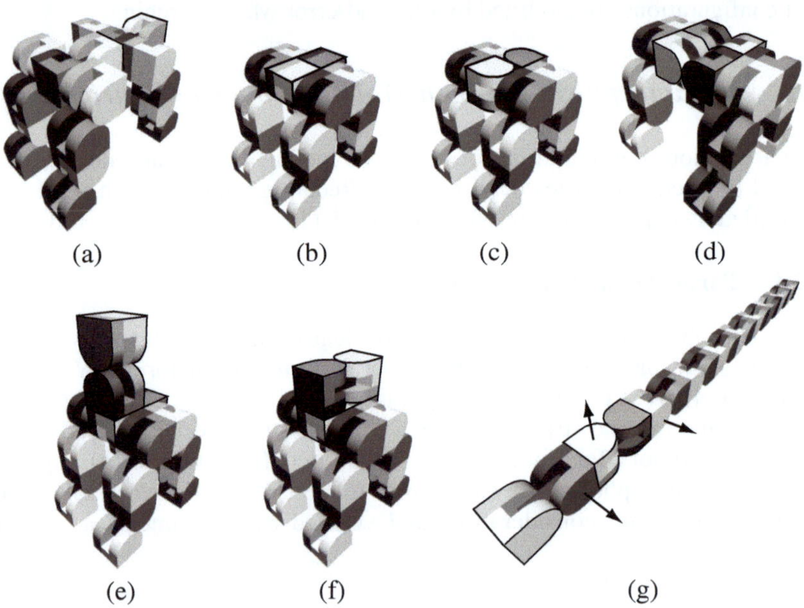

Fig. 7.12 Parallel quadruped form

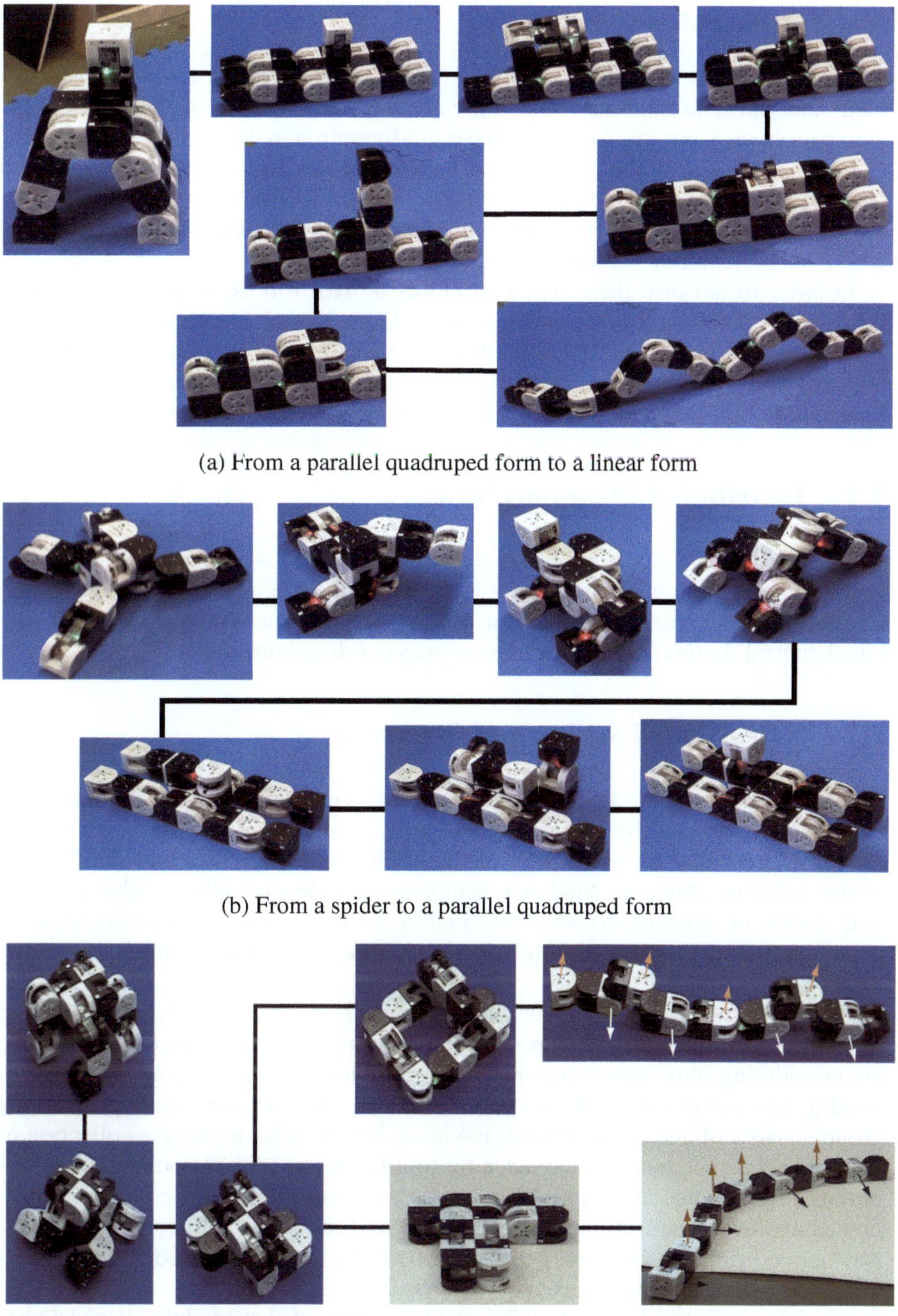

(a) From a parallel quadruped form to a linear form

(b) From a spider to a parallel quadruped form

(c) From a cross-shape quadruped form to a linear form

Fig. 7.13 Metamorphosis experiment

7.2.2.2 Other Metamorphoses

The minimum quadruped form in Fig. 7.11 with four modules has rotational symmetry and plays an important role in the next section. We discuss in Chapter 9 the metamorphosis from this form to a linear configuration in detail.

Fig. 7.13(b) shows the sequence of metamorphosis from the spider form with ten modules to the parallel quadruped form. Fig. 7.13(c) show transitions from the cross-shape quadruped form with eight modules to two types of linear forms, whose axis orientations come in different orders.

In order to perform these metamorphoses on real robots, it is not quite sufficient that a metamorphosis pathway exists. Even if we obtain various paths between two configurations, some of them likely cannot be carried out because of physical constraints or because the whole robot falls due to gravity. Therefore, transient motions between in-lattice states also need to be carefully designed.

7.3 Distributed Metamorphosis

All the examples of metamorphoses we discussed so far were executed under centralized control following procedures planned in advance, not yet considering how metamorphoses could be carried out by the modules autonomously or by distributed control. In this section, we explore ways of distributed control.

7.3.1 Distributed System and Grouping

When the number of modules is large, searching for a metamorphosis procedure is unrealistic in view of the complexity, either by heuristics or by exhaustive search. Even if a procedure is obtained by searching, it needs to be executed step by step in the order as obtained. Modules cannot move independently, or else modules will collide or connections will fail. Such a kind of stepwise execution requires centralized control under a single leader and/or synchronization of the whole system.

Generally speaking, in dealing with a large system, it is effective to divide the system into subsystems and impose a hierarchy on them. For metamorphoses we are considering, this means that modules are divided into groups, motions are defined in terms of groups, and the system is dealt with as a collection of groups. If a group consists of a small number of modules, it is possible to make a collection of possible action sequences in advance by limiting the number of possible configurations and motions, and build procedures by choosing from this collection. Such procedures in the collection need to be executed internally in a centralized manner, but coordination between groups can be done flexibly under distributed control. How then should modules be grouped?

Suppose modules A and B in Fig. 7.14(a) move so that modules B and D become connected as in (b). In order to achieve this when modules are running with distributed control, it is necessary that all four modules, from A to D, work in coordination. As we mentioned earlier, such coordination requires some kind of centralized control, e.g. A sends out a message to the other three modules, and on receiving it the modules C and D stop their motions. By message exchange and coordinated motions like this, the four modules act as a group, and a communication route within the group can be established.

Now, consider that modules outside this group might be in the way of B's motions as E and F in Fig. 7.14(c). In order to perform the motion in (b), it is necessary either to confirm that a module such as E does not exist or to ask E to get out of the way. However, in Fig. 7.14(c), the module E is not connected directly to the group. Module A can send a message to discover the existence of E or to coordinate with E, but for the message that goes outside the group, the distance to the destination is unknown. To ensure the communication and coordination, it is necessary to send a message to all the modules in the system and to coordinate with them. This is no different than centralized control of the entire system.

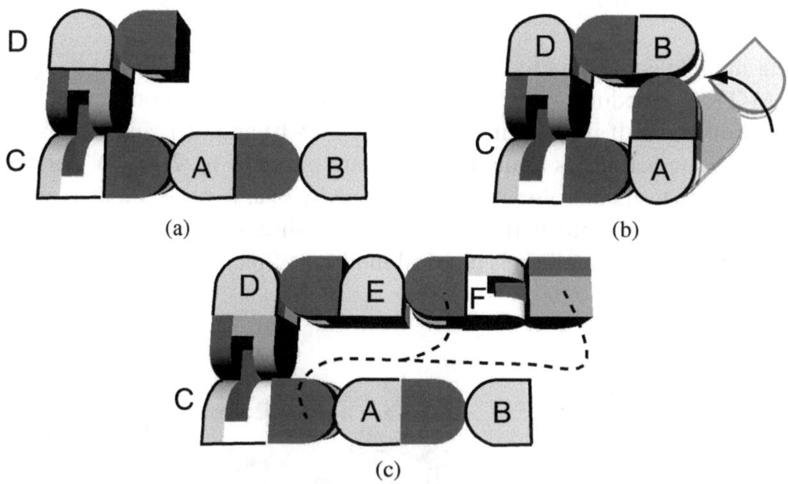

Fig. 7.14 Distributed control problem of M-TRAN. Broken lines indicate that modules are connected via other modules.

We would like to resolve such communication and coordination issues as locally as possible. For example, the modules may be grouped so that when two groups are adjacent, the communication path between them is automatically determined, and thus when module E belongs to a group adjacent to A's group, the module A can discover the existence of E through message exchanges between these two groups only. In fact, the mismatch of the geometric distance and the distance in a connection network is a global problem that cannot be resolved only by grouping; we will discuss this later in Section 7.3.5.2. Still, for the purpose of avoiding collisions, mutual exclusion, and coordination, it is important to try to keep spatially neighboring groups connected so that there is a direct communication path between them.

Two similar methods have been proposed to realize the concept of grouping discussed above: one is what is called a *meta-module*, a group that simulates motions of a virtual module with higher symmetry. This has been proposed for various modular robots including M-TRAN[2]. The other is *regular structures*, proposed for M-TRAN in particular. The idea here is to restrict rather than increase the symmetry of the entire configuration. Modules are grouped into a unit and units are repeated to form a regular structure. We also call this unit a meta-module.

In either method, the use of meta-modules allows a hierarchical control structure. Meta-module motions, executed by coordination of the modules in the meta-module, comprise a lower layer of this hierarchy, while distributed control is realized in upper layers.

7.3.2 Meta-modules Simulating Virtual Modules

A meta-module equivalent to the Crystalline module (Fig. 6.11) can be built from eight M-TRAN modules (Fig. 7.15). This meta-module can connect with its four sides, shrinks in half in two directions (Fig. 7.15(b) and (c)), and moves by a sequence of basic actions as in the fifteen square puzzle (Fig. 7.15(d)). For example, in order to move squares along the arrows one by one, it is sufficient to shrink two neighboring squares into half, and shift these shrunken squares to an adjacent square. When reconfiguration is planned in terms of meta-modules, collision is not

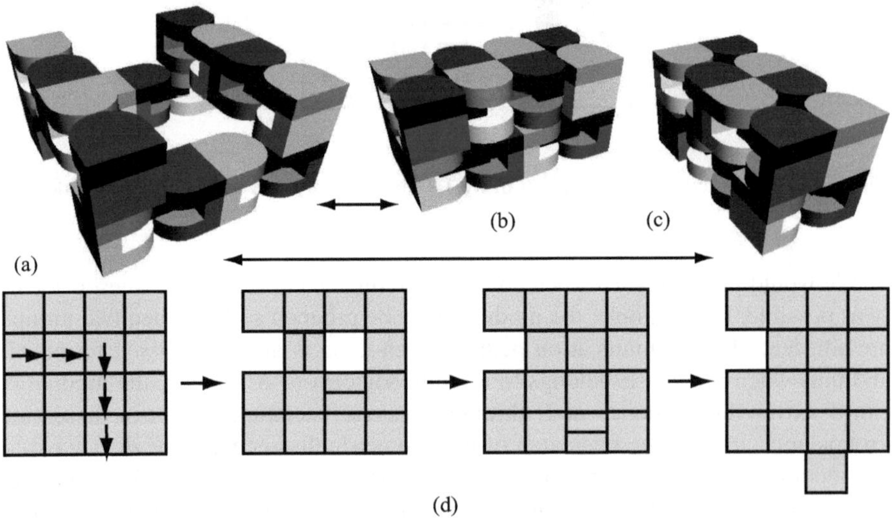

Fig. 7.15 M-TRAN meta-module

[2] The meta-module idea was first proposed in [10, 11].

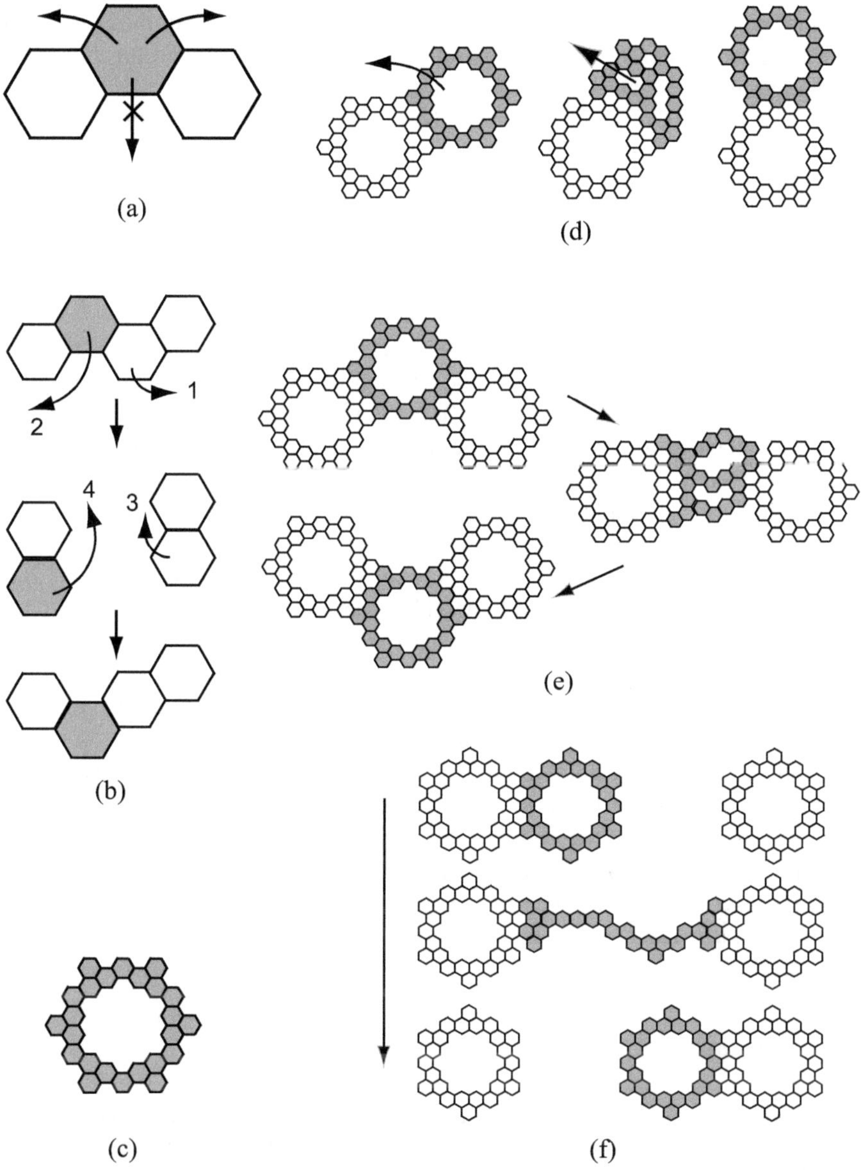

Fig. 7.16 Fractum meta-module

an issue, splitting of the robot is also easily avoided, and meta-modules placed apart from each other can move in parallel. Algorithms developed for the original Crystalline modules can be applied as they are [12].

The above meta-module does not change the connection relation among its components to attain its function, but the next example does. Consider constructing a Fractum-like meta-module using the original Fractum modules. With the original Fractums, in the situation shown in Fig. 7.16(a), the grey module cannot make the downward move. In some cases a workaround procedure may be possible, as shown in (b), but it is preferable that it can be done in any case. The figure (c) is a possible meta-module that allows such a move.

First, let us see if this meta-module has the same motion capability as the original Fractum. Each module moves in the original motions of a Fractum as shown in the picture in the middle of Fig. 7.16(d). Now, if we ignore the intermediate states and simply look at the pictures on the left and right in (d), it seems as if a large hexagon moved like a Fractum module.

By using this meta-module, motions such as passing through a narrow gap (Fig. 7.16(e)) and a jump over a distance (Fig. 7.16(f)) become possible. In the course of these motions, it is possible to maneuver each module so that the moving meta-module stays connected to neighbor meta-modules.

The procedure of passing through a narrow space as in Fig. 7.16 (a) makes the metamorphosis process very adaptable and expands the range of possible configurations. For example, consider a configuration of packed modules (Fig. 7.17, on the left). If the hexagons are original Fractum modules, only those at the corners (indicated by white dots) are movable, while other modules on the edges and the inside cannot move because there will be collisions with adjacent modules. Therefore, transforming the configuration on the left to that on the right requires a complicated procedure. On the other hand, if these hexagons are the meta-modules described above, even the internal hexagons can move as indicated by the arrows, allowing simple metamorphosis from a filled configuration to a ring configuration[3].

7.3.3 Regular Structures

The above conception of a meta-module is purely geometric, and as seen from the examples above, the number of modules required for a meta-module is quite large. The idea of a *regular structure* involves repeated use of a unit made up of a small number of modules according to a principle as in Fig. 7.18(a). This unit is also called a meta-module. If individual modules in a meta-module make motions independently as (b), and if we ignore the intermediate processes and consider only the beginning and the end of the motions (Fig. 7.18(c)), it seems as if in this process a meta-module moves while retaining its outline structure.

---

[3] A meta-module with similar functionality called a scaffold, which is a combination of virtual cubic modules, also has been suggested [13].

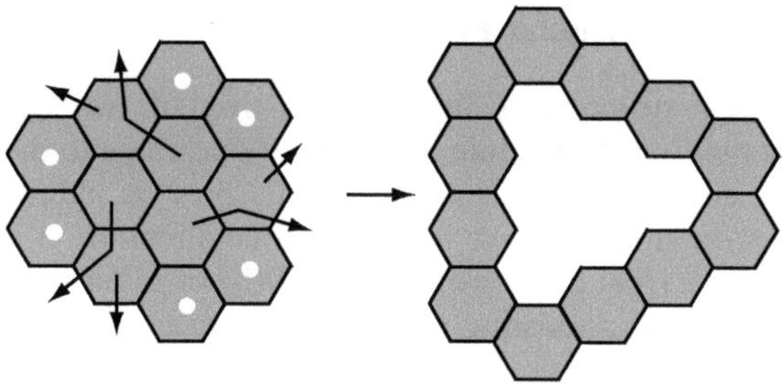

Fig. 7.17 Advantage of meta-modules

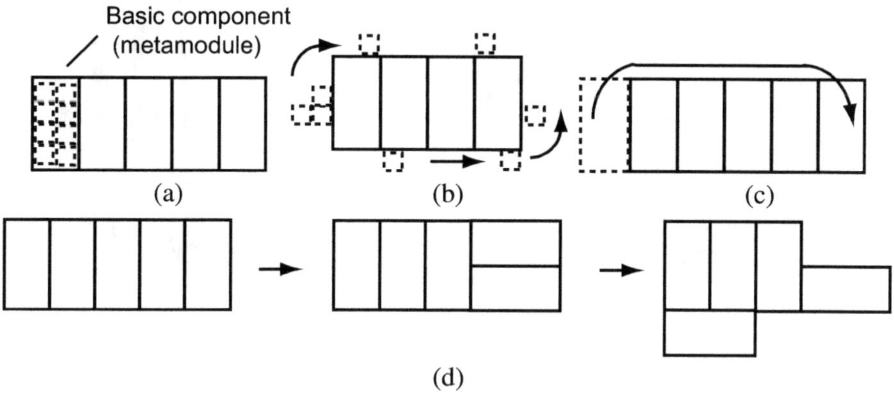

Fig. 7.18 Motions of a regular structure

The meta-modules in the previous section meet the following constraints in order that they can simulate virtual modules:

1. Component modules move only inside the virtual module
2. A virtual module moves as a unit to adjacent positions only
3. A meta-module is always composed of the same component modules.

On the other hand, meta-modules in regular structures are not constrained by these conditions. For instance, (b) and (c) in Fig. 7.18 are cases where Constraints 1 and 2 are not satisfied. It is also possible to obtain (d) after the motion in (c) if we do not need to meet Constraint 3.

Although there are differences in conception, there is not much point in distinguishing these two kinds of meta-modules strictly. In either case, metamorphosis of the entire system by self-organization is facilitated by making meta-modules the functional units of the system.

Various M-TRAN regular structures have been proposed so far: linear, planar, and three dimensional [14, 15].

7.3.3.1 Type-I Linear Regular Form

The simplest regular structure is linear, consisting of two chains of modules (Fig. 7.19(a)). By moving the module at an end of one chain to the other end while keeping the other chain fixed, and repeating this process alternately on the two chains, the entire structure can move in a line in one direction. It is also possible to bend the structure in a right angle as in Fig. 7.19(b) or to turn it sideways using a device called a converter, consisting of two modules, as shown in (c). This structure, including the converter, is called a Type-I linear regular form.

(a) Basic structure

(b) Bending at a right angle

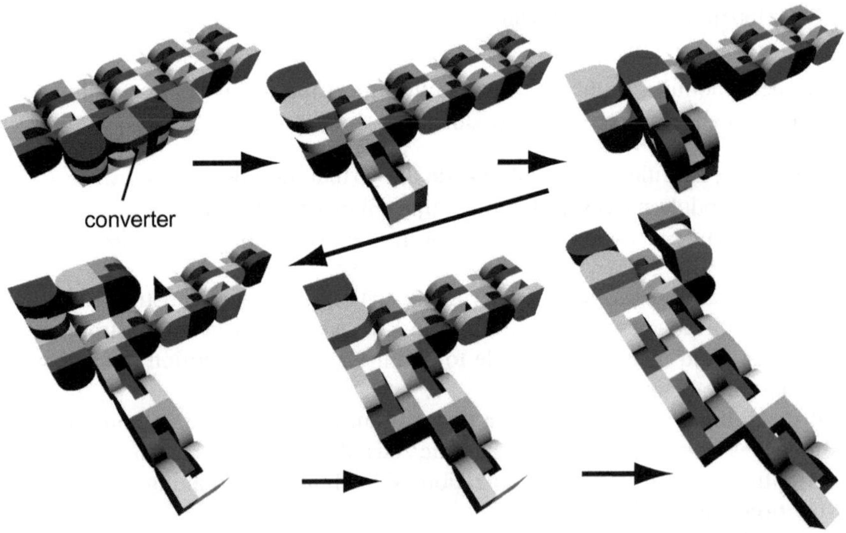

(c) Changing direction.

Fig. 7.19 Type-I linear regular form

7.3.3.2 Type-II Linear Regular Form

In the structure shown in Fig. 7.20, which is a regular linear sequence of cubes each consisting of four modules, each cube can be moved from the tail to the head. Again, by adding a converter consisting of two modules, the structure can also be made to turn or branch up, down, left or right at a right angle. This structure is called a Type-II linear regular form.

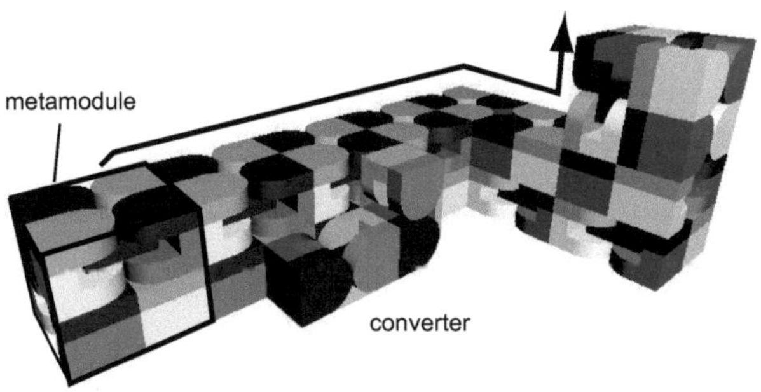

Fig. 7.20 Type-II linear regular form

7.3.3.3 Planar Regular Form

Fig. 7.21 shows a planar structure which is built by a repetition of crosses (the minimum quadruped form in Fig. 7.11) each consisting of four modules. Transformation by distributed control will be detailed in Sections 7.3.4 and 7.3.5.

7.3.3.4 Three Dimensional Regular Forms

Using the same cross-shaped four-module groups as building blocks, it is also possible to obtain various three dimensional regular structures (Fig. 7.22). Procedures can be developed for moving these meta-modules to a neighboring lattice point.

7.3.3.5 Cluster Flow

Meta-module structures can move themselves like an amoeba by repeating local motions. Motion of meta-modules in parallel is called *cluster flow*, because it looks as if many meta-modules are flowing.

Let us consider a cluster flow of the Type-I structure in one direction. There are two kinds of this structure, shown in (a) and (b) of Fig. 7.23. Also, there are two possible ways for modules to move locally so that the whole structure moves forward. One is to carry a module at one end to the other end as shown in (c), and the other is to successively move two modules in the forward direction, thus shifting the resulting gap to the rear as shown in (d). In particular, algorithms for (d) are simpler in that each pair of modules moves forward depending on the existence

Fig. 7.21 Planar regular form

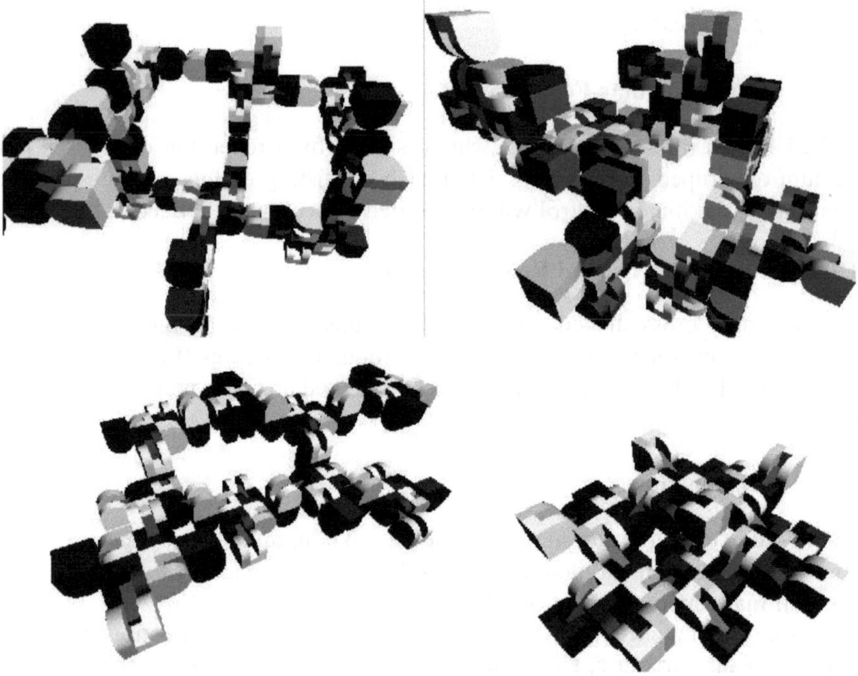

Fig. 7.22 Three-dimensional regular form

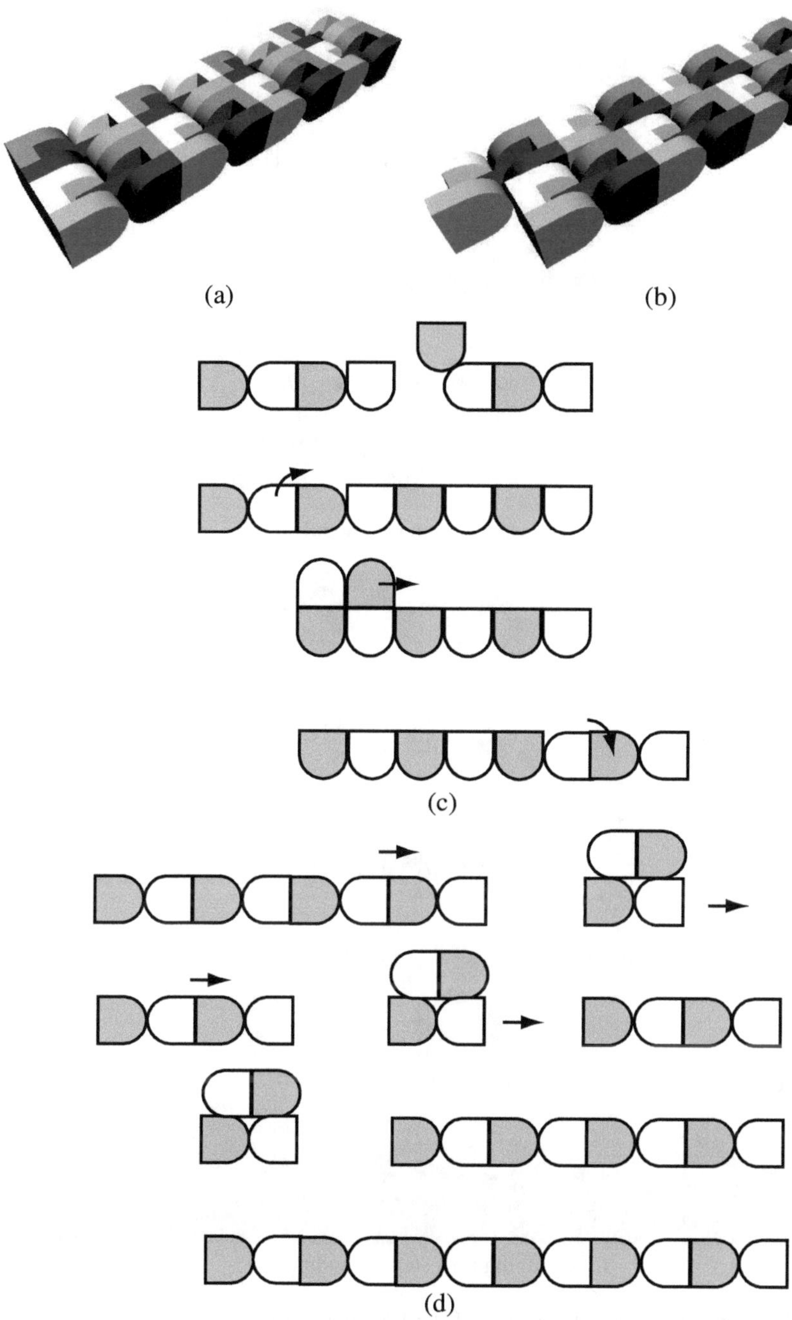

Fig. 7.23 Type-I linear form and forward motion

(a) Forward motion (16 modules)

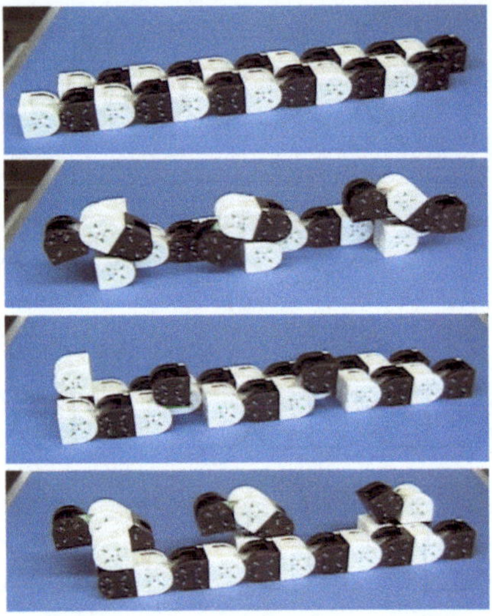

(b) Forward motion (12 modules)

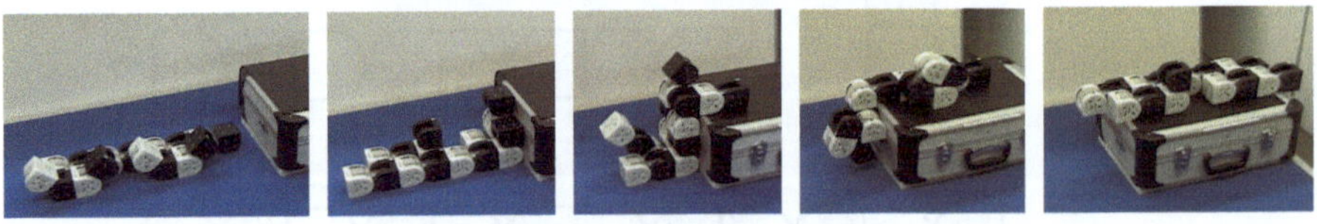

(c) Climbing up a step (8 modules)

Fig. 7.24 Experiments of cluster flow with Type-I regular form and status of modules to their front, rear and side. Cooperation with distant modules is unnecessary.

Fig. 7.24 (a) and (b) show experiments achieving different kinds of forward motion. For each of these motions, the same program is used regardless of the number of modules in the structure. Fig. 7.24 (c) shows a motion of climbing up a step [2].

7.3.4 Motions of Planar Regular Structures

In this section we take a closer look at motions and control of planar regular structures and attempts to make them into distributed autonomous systems.

7.3.4.1 Basic Motions

We here consider three types of meta-modules which form a planar regular structure as shown in Fig. 7.25(a). The motion principle is described in Fig. 7.25(b). First, two modules in an L-shape at an edge of the planar structure are lifted onto the plane, where the two modules are serially connected and their joint axes are made vertical. A pair of modules in this configuration is called a *walker pair* and the above process is called *walker generation.* A walker pair can traverse a plane with a motion similar to pivoting if it can connect to the plane underneath it. When

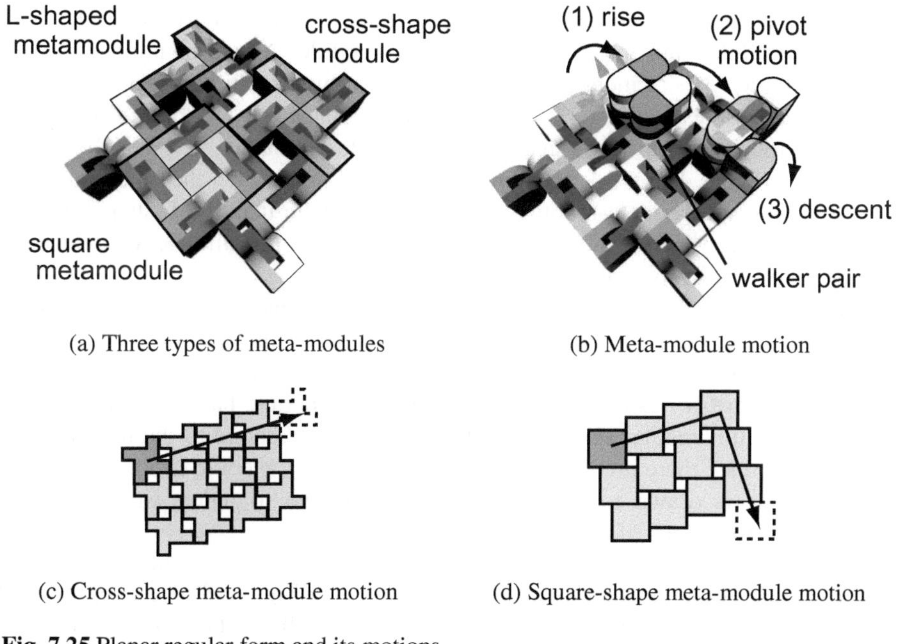

(a) Three types of meta-modules

(b) Meta-module motion

(c) Cross-shape meta-module motion

(d) Square-shape meta-module motion

Fig. 7.25 Planar regular form and its motions

a walker pair reaches another edge of the structure, a process which is the reverse of walker generation is executed, and the pair reverts to be a part of the regular structure, changing the shape of the configuration as a whole. In this way, an L-shaped meta-module can move from one edge to another, and by repeating this process twice, a meta-module of a cross or square shape can move similarly (Fig. 7.25(c), (d)).

There are two types of walker generation, shown in Fig. 7.26(a) and (b), depending on the relations between the L-shaped pair and modules around it. In either case, coordination with four modules (a, b, c and d in the figure) is necessary. In order for the walker pair to move by pivoting, it has to connect to the planar structure. However, normally all the modules in a planar structure have their connection faces facing each other in order for the structure to be connected as tightly as possible, as shown in Fig. 7.25(a). Therefore, a module in charge is required to turn its connection face upward to support the walker pair. We call such an upward face an *anchor*. An anchor is formed by the same motion as in Fig. 7.7(b) so that there is no collision with neighboring modules (Fig. 7.27).

The basic processes of a planar regular structure are walker generation, anchor formation, reversals of these, and pivoting of a walker pair. By carrying out these processes in an appropriate order, self-reconfiguration of a planar regular structure is achieved.

7.3.4.2 Tile Model

For self-reconfiguration by distributed control, we need metamorphosis rules both for meta-modules and for modules in a meta-module. As we mentioned earlier, there are three types of meta-modules possible for this structure: L-shaped consisting of two modules, and cross-shaped and square-shaped consisting of four modules. Obviously, since the L-shaped meta-module is smaller, it gives the whole structure a higher degree of freedom in shape and motion. On the other hand, the cross and the square meta-modules are easier to handle than the L-shape because of their symmetry. Since the cross and the square are quite similar, we focus here on the cross shaped meta-modules.

We now define an abstract model to denote a regular structure and its metamorphosis [16]. Given a planar regular structure, we can draw lines passing the points where four modules meet as shown in Fig. 7.28(a) and get a square grid, so that each square corresponds to one module. Each module has connections at its four faces in this regular structure, and these faces correspond to the four sides of the square. Since there are two possible placements of a module inside a square, a diagonal line is drawn to divide the square into two triangles, each of which corresponds to a block of the module.

Fig. 7.28(b) is a description of (a) using this abstract model. The cross-shaped meta-module A and anchors indicated with circles in Fig. 7.28(a) are represented by the two-by-two squares and the gray triangles in (b), respectively. In (b), the

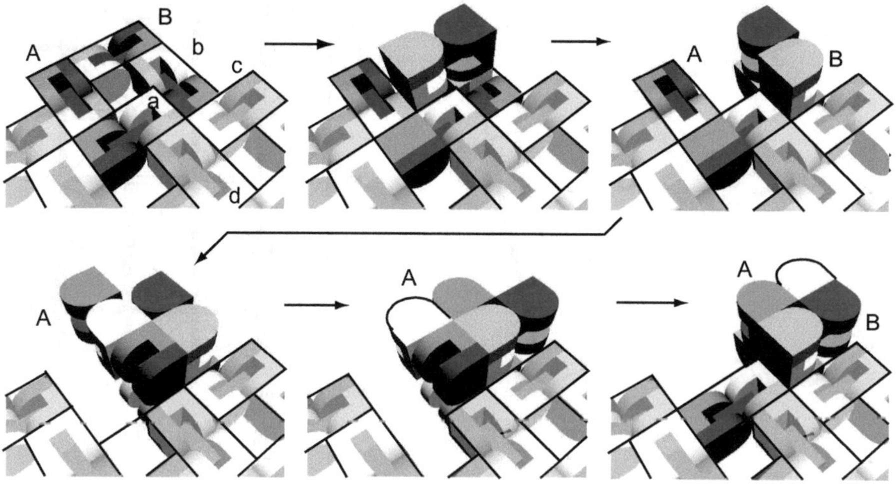

(a) Procedure 1

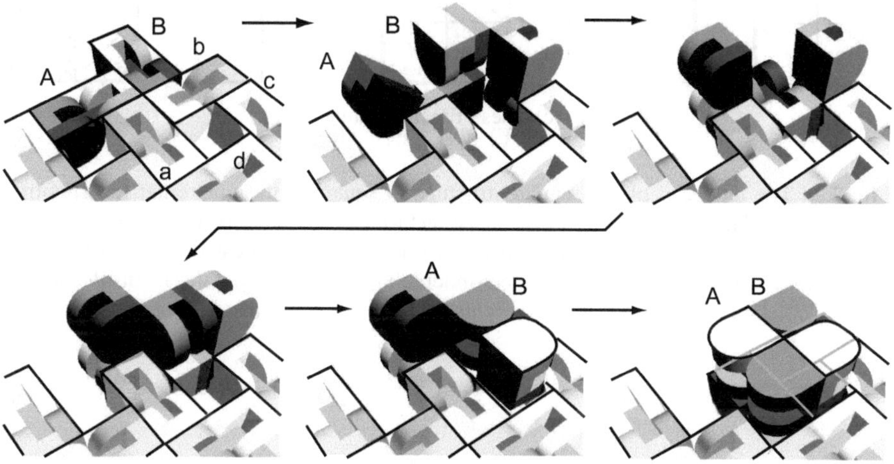

(b) Procedure 2

Fig. 7.26 Walker pair generation

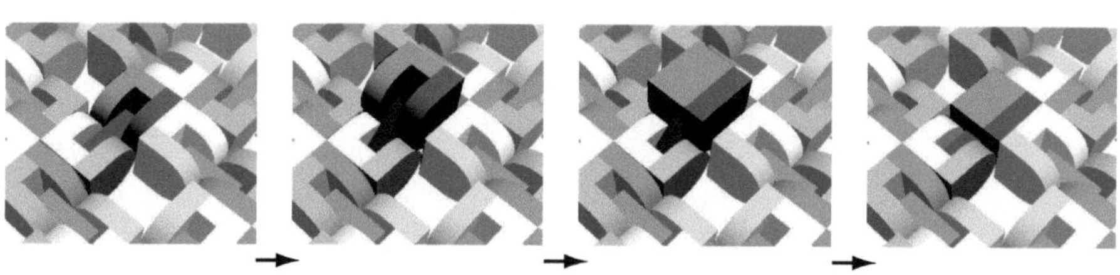

Fig. 7.27 Anchor formation

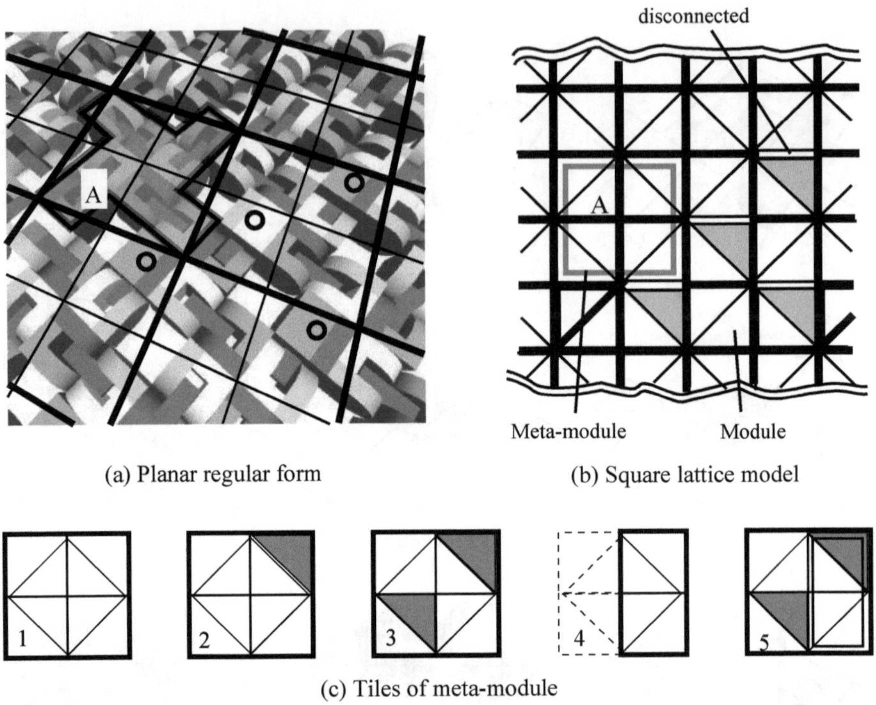

Fig. 7.28 Tile model of planar regular structure. A square in the square lattice (b) corresponds to a module in (a), while a cross-shape meta-module (A) in (a) corresponds to the 2 x 2 square (A) in (b). The anchors marked with circles in (a) are indicated by grey triangles in (b). In (c), local configurations of a meta-module are expressed as patterns of tiles and by discrete numbers. The tile 5 in (c) represents the state where a walker pair is connected to a tile 3.

connections are indicated by solid lines, so that in contrast the absence of lines indicates the blocks of the modules forming anchors that are disconnected from their neighbors. Actually, we omit to draw such lines hereafter, because given the dark triangle representing an anchor, the disconnected side is automatically determined.

A planar regular structure can be considered as a plane filled with tiles of 2×2 squares, which we call a *tile model.* Tiles may have various patterns as in Fig. 7.28(c). The tile 4 in (c) represents the rest of a meta-module after a half of it is transformed to a walker pair. A walker pair can be indicated by distinguishing the tile by some attribute marker.

Since the tile model is only an abstraction, various control rules and procedures have to be provided. In particular, it is important to guarantee collision avoidance and entire structure's connectivity under distributed metamorphoses. Below, we summarize procedures:

1. **Anchor formation:** In the model, anchor formation is represented by change of the color of triangles. Before the walker pair makes a move, there must be an anchor at the position where the walker's end lands on the plane. When creating a necessary anchor using the process in Fig. 7.27, coordination of tiles in the neighborhood is necessary to preserve connectivity.
2. **Walker generation:** When generating a walker pair, cooperation with neighboring meta-modules is indispensible. Procedures for walker generation differ depending on the patterns of tiles. Fig. 7.29 is an example.
3. **Walker pair movement:** When moving the cross-shaped meta-module from one edge to another, two walker pairs are generated and move separately from each other. In order to make these two arrive at the same destination, the walker pairs should be guided properly so that the first one leaves marks on the anchors on its path, and the second one follows the marks.
4. **Communication within a meta-module:** The modules communicate with each other through their connections. It is desirable that the four modules forming a meta-module communicate within the meta-module only through their connections. For example, all component modules are connected within meta-modules in all of the four meta-module tiles in Fig. 7.28(c) except the fourth.

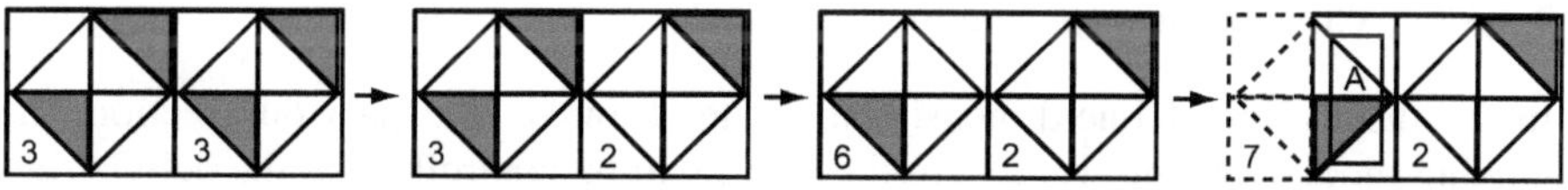

Fig. 7.29 Walker pair generation

7.3.4.3 Cellular Automaton

If we regard the tiles in this model as basic components of a distributed system, the whole system can be formalized as a cellular automaton. As shown in Fig. 7.30, tiles correspond to cells, and the state of a cell is 0 if there is no tile, otherwise tile pattern number as in Fig. 7.28(c), Fig. 7.29, and etc. Together with a set of transition rules that determine the next state of a cell depending on its current state and neighbor cells' states, we obtain a cellular automaton (Fig. 7.30).

This is an asynchronous cellular automaton, in which cells do not make state transitions at the same discrete times. In addition, there is no tile or module that corresponds to a cell with zero state, so there is no information processing facility for such cells.

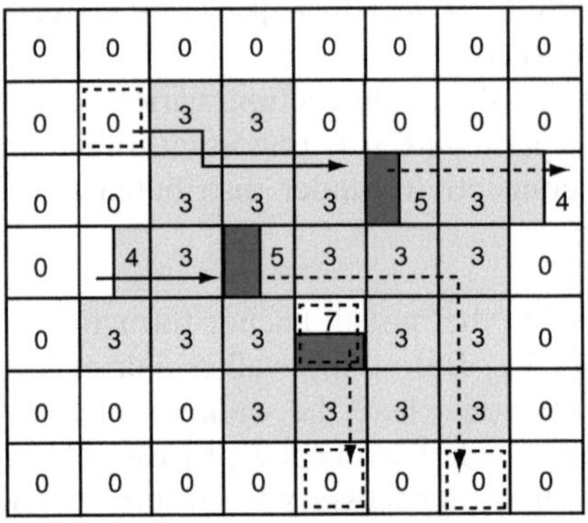

Fig. 7.30 Cellular automaton model and its behavior. Walker pairs in motion are indicated by dark rectangles for descriptive purposes.

7.3.5 Distributed Metamorphosis by the Cellular Automaton Model

Now that we have modeled the M-TRAN planar structure as a cellular automaton, we can run a distributed metamorphosis by setting appropriate rules of the cellular automaton. The actual rules and algorithms depend on the tasks to be completed. There have been various studies in which computer simulations have been made using ideal asynchronous cellular automata models. Various tasks have been considered such as a cluster flow, transformations to target configurations, and reaching target points [17, 18].

7.3.5.1 Cluster Flow on Planar Structure

In the discussion of one-dimensional cluster flows in the previous section, the moving modules themselves were the objects to be controlled. In the cellular automaton model in this section, however, a moving walker pair is regarded as an attribute of the cell, and the object to be controlled is the cells that are fixed in the space. Such difference in perspectives corresponds to the difference between Euler and Lagrange formulations in dealing with fluid dynamics. In the cell space, what we need to do is to generate a flow field in the cell space.

The problems of forming target configurations and avoiding obstacles can be expressed by setting a vector field in the cell space as in Fig. 7.31. Here we do not consider how to describe the target configuration. A vector field can be generated by a distributed system described in Chapter 4, which simulates gradient generation by diffusion dynamics.

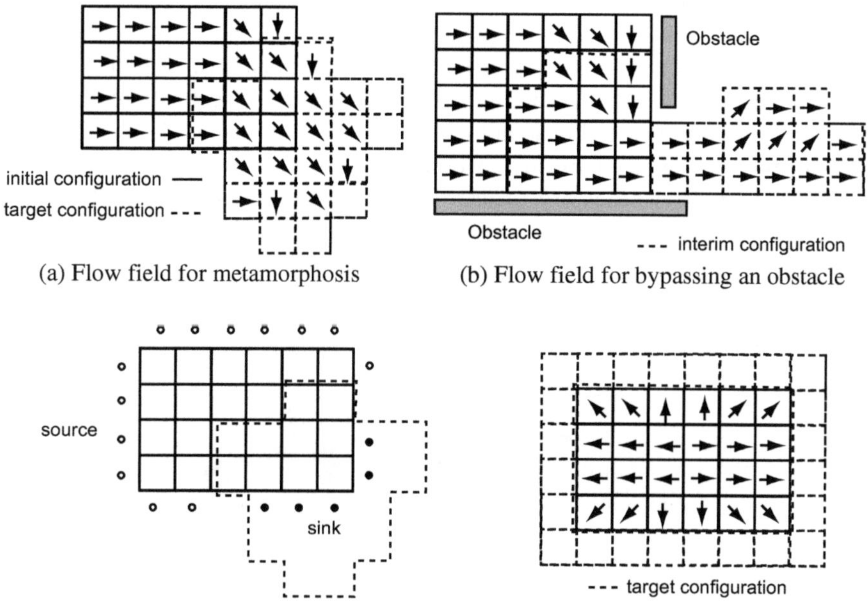

(a) Flow field for metamorphosis (b) Flow field for bypassing an obstacle

(c) Sink and source allocation for diffusion field (d) Flow field for cavity formation

Fig. 7.31 Metamorphosis according to a flow field

For example, a vector field can create a gradient in a diffusion field by placing sources and sinks right next to cells on the edge of the structure (Fig. 7.31(c)). If each module is capable of detecting obstacles, it is possible to create flows that follow the contour of the environment (Fig. 7.31(b)). Walker generation starts at the sources in Fig. 7.31(c), walker pairs move in the direction of the vector field, and then at the sinks, new cells are generated.

There is a certain freedom in how to define the vector field. In cases like Fig. 7.31(b), after the whole goes through the narrow gap by obstacles, the shape of structure can be controlled rather freely. An example of such a method is to introduce surface tension along the edge of the area which makes the area circular [19]. A ring shape can also be made by placing a source in the center, and generating a vector field that sends the cells outwards as in Fig. 7.31(d).

Fig. 7.32 shows results of simple cluster flow experiments of cells arranged in rectangles moving in one direction. With a unidirectional flow, collision avoidance is much simpler. By installing the same program in all the modules, flows of clusters consisting of $1 \times n$ cells (for $n = 2, ..., 5$), 2×2 cells, and 3×2 cells were made [2].

7.3.5.2 Collision Avoidance, Deadlock Avoidance, and Global Consensus

Modules in motion have to avoid colliding with each other. A natural way to do so is to assign to each a certain area into which others cannot enter. This is simple mutual exclusion by maintaining distance between moving objects, but this may cause local deadlocks and various global problems.

Fig. 7.33 shows two examples of local deadlock situations. In both cases, the vector fields have vortices. As we explained in Chapter 4, vector fields generated from gradients of scalar fields do not make vortices in principle. As a cell space is discrete, when an area of exclusion is assigned to each module, deadlocks are prone to occur. It is necessary to use distributed algorithms for avoiding deadlocks, both at the level of the meta-module and at the level of the individual modules.

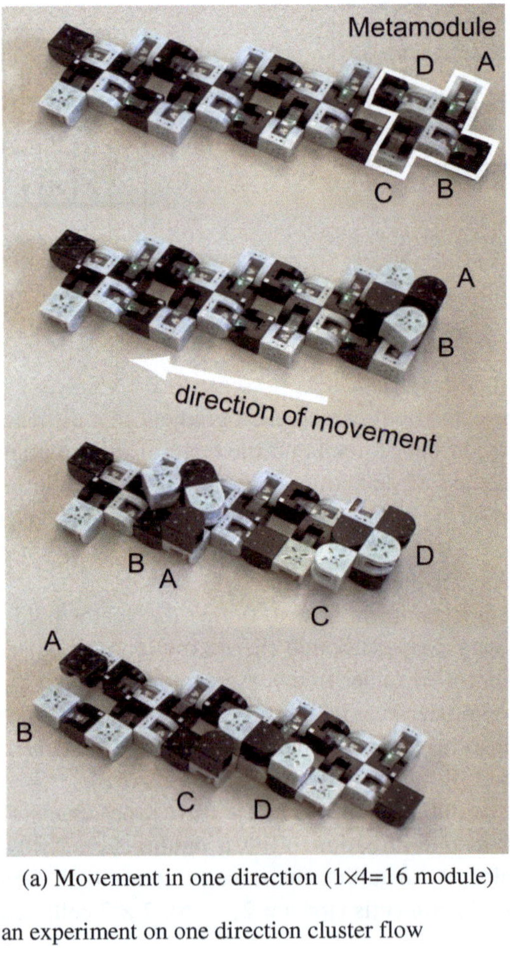

(a) Movement in one direction (1×4=16 module)

Fig. 7.32 Result of an experiment on one direction cluster flow

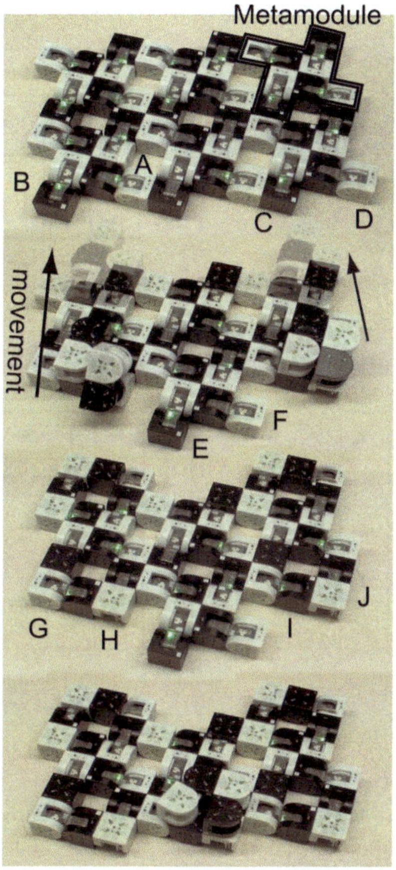

(b) Movement in one direction (3×2=24 module)

Fig. 7.32 *(continued)*

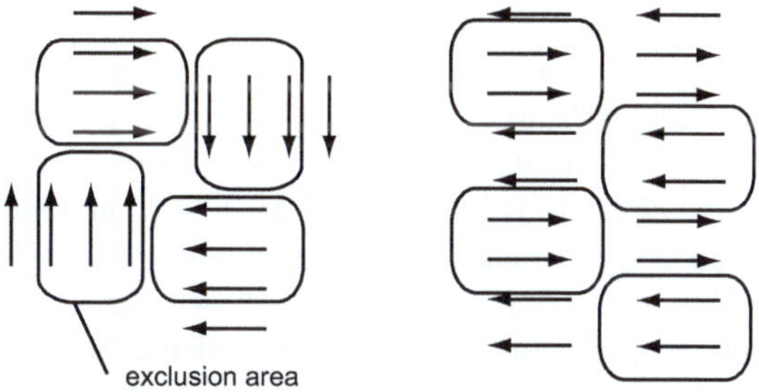

Fig. 7.33 Local deadlock

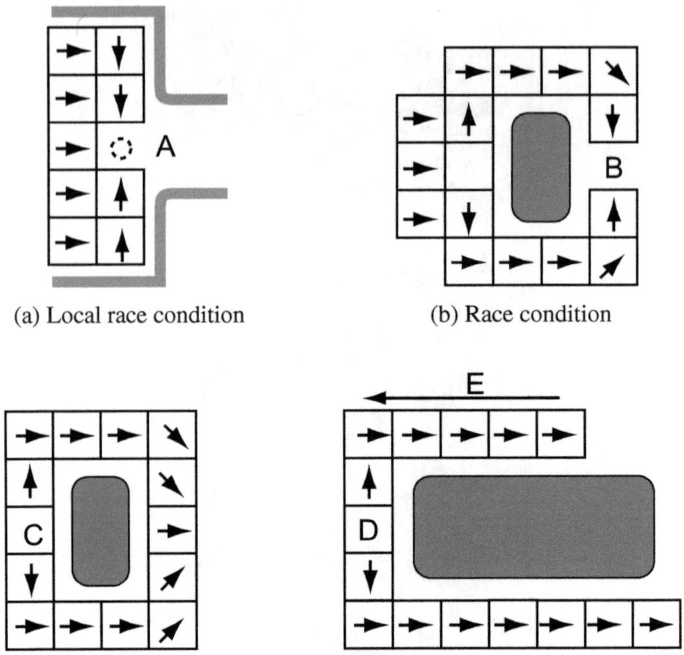

(a) Local race condition (b) Race condition

(c) Connectivity (d) Deadlock

Fig. 7.34 Race conditions and global issues

Fig. 7.34(a) shows a race condition where several nearby cells are about to proceed to the cell A. This situation is potentially a global bottleneck, but forward motion is still possible by local coordination among neighboring cells. However, such coordination is not possible in the case shown in Fig. 7.34(b), where a similar condition occurs after cells detour in two routes around an obstacle. In this case, it is impossible even to recognize the possibility of collision, let alone coordinate motions, only by local information exchange.

This problem is due to the lack of global information such as the shape and the size of an obstacle. Similarly in Fig. 7.34(c), whether separation at C is possible without breaking the connectivity of the whole is a similar global issue. Moreover, in the case where the obstacle is large as in Fig. 7.34(d) and the two branches do not join up, it is necessary to make a global decision on whether to separate at D or to reverse direction in one of the branches as shown in E.

Some of the above problems are caused by mismatch between the physical neighborhood and communication neighborhood of each module or meta-module. These problems can be alleviated by improvement of hardware. For instance, if a module can directly recognize the modules in a slightly extended neighborhood using sensors, the possibility of collision in Fig. 7.34(b) can be foreseen by the modules themselves. Race conditions can be avoided if unconnected modules can communicate directly with each other.

7.3.5.3 Porous Structure

When a structure with a hole is the target configuration to be formed, such global race conditions do not occur. When building the configuration in Fig. 7.35(c) starting from Fig. 7.35(a), we can specify the final length of two branches as in Fig. 7.35(b) and avoid the above problems. This is equivalent to centralized decision making.

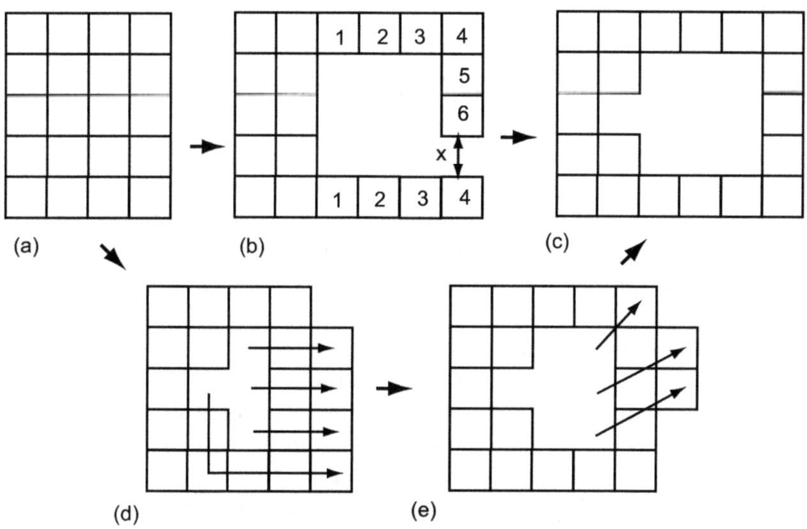

Fig. 7.35 Ways to form a configuration with a cavity

In an actual situation, however, the above process of extending branches may fail. The size of the gap indicated by x in Fig. 7.35(b) is exactly one cell in the ideal model, but in real physical structures errors are inevitable. Longer branches will have larger error, and will cause collision or connection failure.

On the other hand, the same structure can be formed by creating and expanding an internal opening as in the sequence of Fig. 7.35 from (a), via (d) and (e) to (c). This process does not require joining of long branches, and errors are expected to be much smaller.

If we have appropriate lattice type modules, the above method can be extended to form a porous structure, which could be stress-adaptive, transforming itself according to the stress applied, if modules are equipped with stress sensors (See Section 6.4.7) [20].

7.4 Various Metamorphoses

The advantage of meta-modules is that they extend the symmetry and functions of a single module. One might think that it would then be better to build a module which is functionally equivalent to a meta-module from the start. In reality, making a choice between complex and highly functional modules and meta-modules which are combinations of simple modules is not an issue of principles but rather of hardware realizability.

Another advantage of meta-modules is that they can be ungrouped so that each module moves independently when necessary. A meta-module structure has the potential to detach modules to generate a variety of smaller robots, or to merge detached modules into it.

7.4.1 Generation of Robots from Regular Structures

A mobile robot can be generated from a regular structure by transforming a part of the structure and then detaching the modified part from the body. Fig. 7.36 shows the process of generating a parallel quadruped robot from a type-II regular structure, by taking ten modules out of three meta-modules at its end. By adding two more meta-modules to the remaining two modules, we can generate another quadruped robot, and in this way it is possible to keep generating robots while consuming modules from the structure [21].

Since the two dimensional planar regular structure is made up of minimum quadruped robots, simply detaching a meta-module makes a walking robot. Similarly, various robots as in Fig. 7.11 can be made from the regular structures and can be transformed to others. This can be regarded as a universal assembly machine that uses a regular structure instead of the supporting plane in Fig. 7.9.

7.4.2 Docking and Merging

Docking and merging are the processes reversing generation and detachment. However, since detachment is an irreversible process, it is not guaranteed that docking is possible. Modular robots are designed with the conditions that all parts are always connected and that modules usually cannot detect and communicate with separated modules. In order for separate modules to become connected, a means for measuring positions of others and communication is required.

In the case of the prototype module M-TRAN III (see Chapter 9), each module can wirelessly communicate with the host computer using Bluetooth devices. Infrared communication is also available between unconnected modules over short distances. Moreover, a camera module compatible with a passive block of an M-TRAN module was developed (Fig. 9.11). These devices were used for docking.

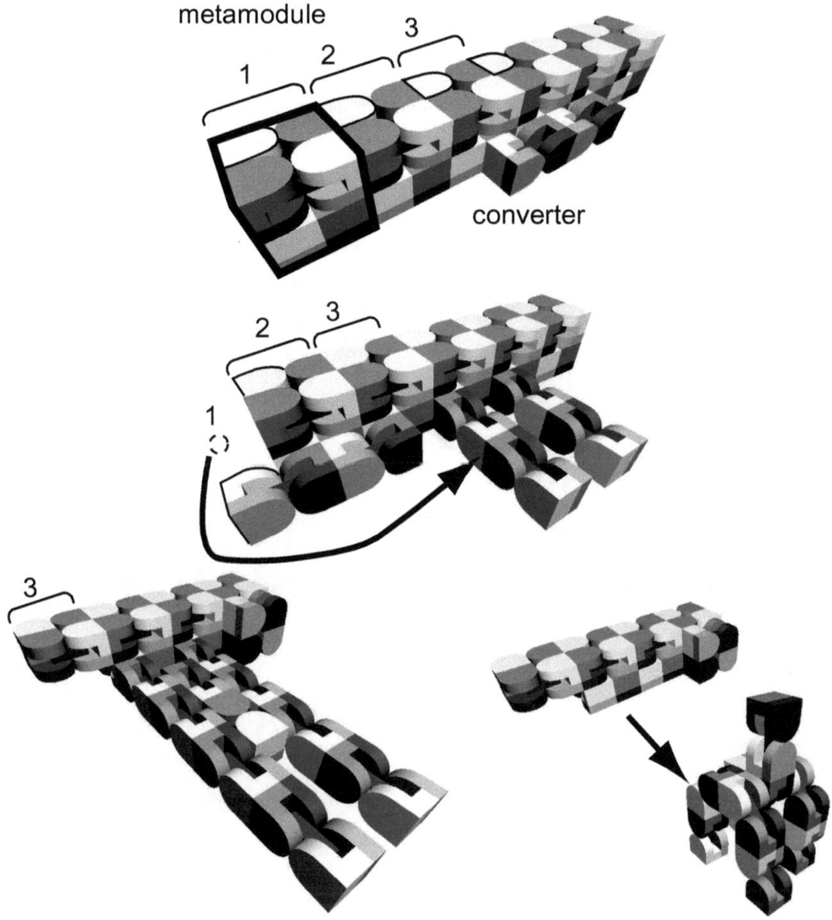

Fig. 7.36 Generation of a walking robot from a type-II linear regular form

Fig. 7.37 shows an experiment in which a mobile robot approaches a Type-I linear regular structure and becomes attached and embedded into the structure. A camera module placed on top of the type-I structure is used for guiding the moving robot to the part called a docking port (Fig. 7.37(a)). Position measurement and guiding of the robot using the camera are both carried out by the host computer, and control commands are transmitted via Bluetooth.

When an end of the mobile robot reaches the docking port, it is clamped and aligned by the docking port (Fig. 7.37(b)). This mechanism enables docking without fine feedback control[4]. After the docking comes a metamorphosis process of embedding the three modules comprising the robot into the type-I linear

[4] In similar docking experiments in [23], alignment errors were absorbed by magnetic connection.

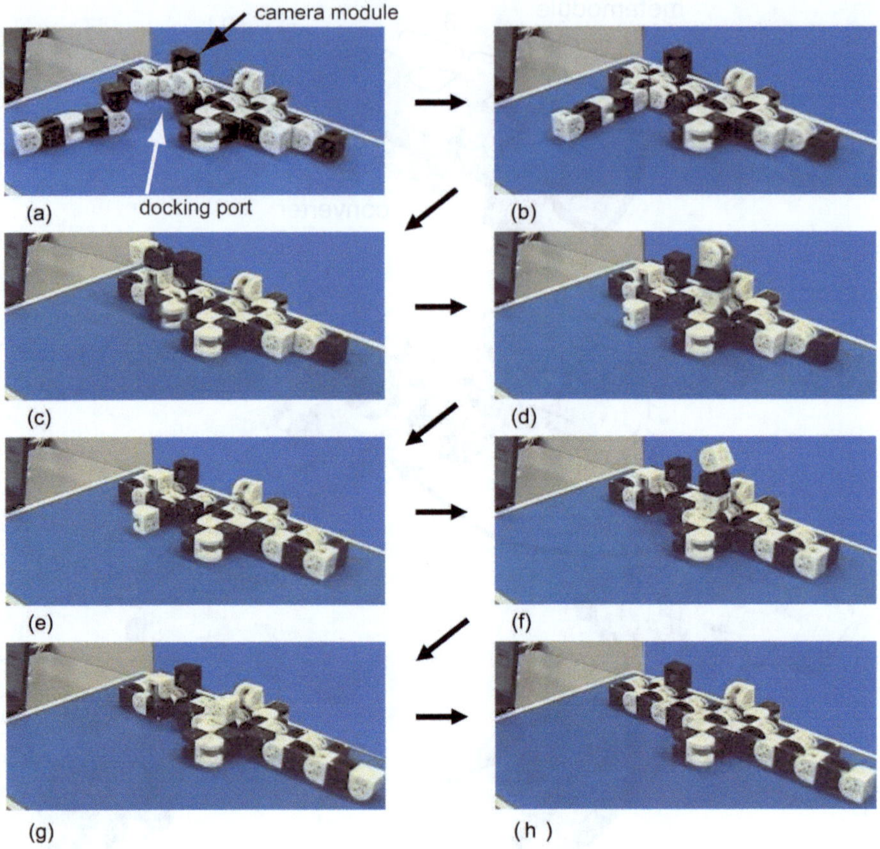

Fig. 7.37 Merging and connecting using a camera module

structure (Fig. 7.37(c) to (h)). In this experiment, all steps from approach to embedding are executed automatically [22].

7.4.3 Self-replication

Another possible goal to be achieved by the robot motions described so far is construction of a copy of the robot by itself, i.e., self-replication [24]. Consider a situation in which many separate modules are scattered on a flat plane, and a robot made of modules gathers and assembles them to make a copy of itself. Instead of considering a walking robot as a parent, we first construct a universal assembler which makes a serial robot of any length. For example, the robot made of nine modules in Fig. 7.38(a) can collect modules in its proximity, adjust the position and orientation of each module (Fig. 7.38(b)) and assemble them to a serial structure by successively putting each on an end of the existing linear structure (c). As this robot can be transformed from a serial robot (d), assembling this serial robot is indeed self-replication.

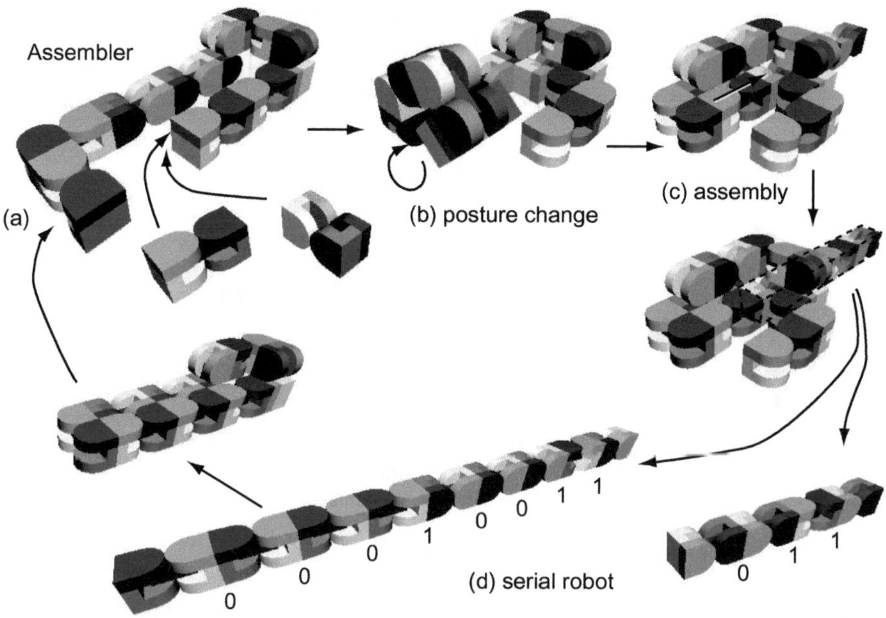

Fig. 7.38 Self-replication concept. Any length of serial structures can be assembled by the assembler in (a). A serial configuration is coded by a binary number, each bit of which indicates whether each neighboring pair is aligned in the same posture or have opposing postures. The serial robot compatible to the assembler is coded as 00010011.

If this serial robot metamorphoses to a walking robot, the whole process may resemble a biological process in which a serial structure (DNA) duplicates itself and grows into an individual. Of course, this is far different from biological self-replication, because ingredients used by natural self-replications are rather passive compared to those in our structure and M-TRAN modules are too active and intelligent to be mere ingredients.

7.5 M-TRAN Colony

In this chapter, we discussed various metamorphoses by M-TRAN modules, most of which were verified by experiments. These metamorphoses, cluster flow movements, detachments, and assemblies are summarized in Fig. 7.39. If these motions are executed autonomously, the robot will be able to continue its task in complex and unknown environments by changing its configuration adaptively. For example, modules will go through a narrow gap in a linear form, travel quickly over a flat surface by walking in a quadruped form, and go over a step in a larger form. Such a capability is suited for searching activities in places inaccessible by humans.

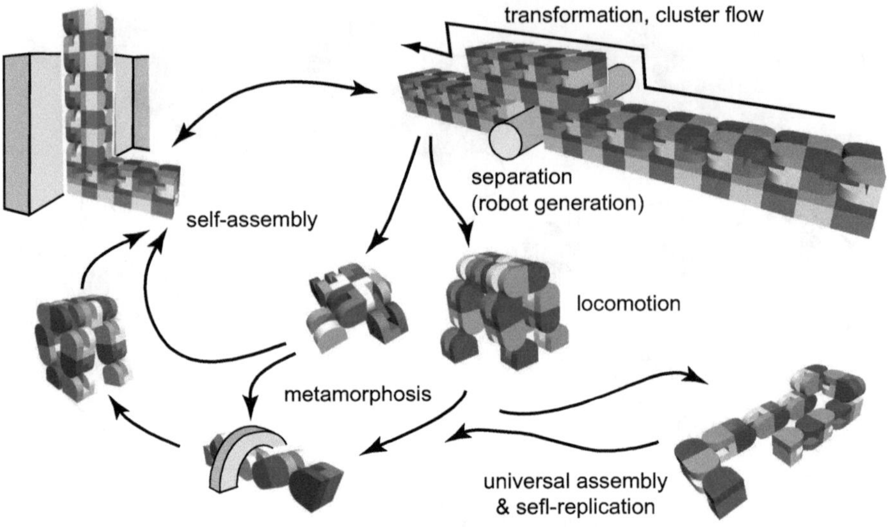

Fig. 7.39 Life cycle model of an M-TRAN colony

The autonomous behavior shown in Fig. 7.39 is similar to that of biological organisms, particularly to the life cycle of slime molds in Fig. 2.11. This resemblance is not surprising, given that this robot behavior was inspired by amoeba movement, metamorphoses of living creatures, and cooperative behavior of ants.

References

[1] Murata, S., Yoshida, E., Kamimura, A., Kurokawa, H., Tomita, K., Kokaji, S.: M-TRAN: Self-Reconfigurable Modular Robotic System. IEEE/ASME Trans. Mechatron 7(4), 431–441 (2002)

[2] Kurokawa, H., Tomita, K., Kamimura, A., Kokaji, S., Hasuo, T., Murata, S.: Distributed Self-reconfiguration of M-TRAN III Modular Robotic System. Intl. J. Robot. Res. 27(3-4), 373–386 (2008)

[3] Kurokawa, H., et al.: A Three-Dimensional Self-Reconfigurable System. Adv. Robot. 13(6), 591–602 (2000)

[4] Chen, I.M., Burdick, J.: Enumerating the non-isomorphic assembly configurations of a modular robotic system. Int. J. Robot. Res. 17(7), 702–719 (1996)

[5] Park, M., Chitta, S., Teichman, A., Yim, M.: Automatic Configuration Recognition Methods in Modular Robots. Int. J. Robot. Res. 27(3-4), 403–421 (2008)

[6] Asadpour, M., Sproewitz, A., Billard, A., Dillenbourg, P., Ijspeert, A.: Graph signature for self-reconfiguration planning. In: IEEE/RSJ Int. Conf. Intell. Robot. Syst. (IROS), pp. 863–869 (2008)

[7] Hou, F., Shen, W.M.: On the Complexity of Optimal Reconfiguration Planning for Modular Reconfigurable Robots. In: Proc. IEEE Int. Conf. Robot. Autom. (2010)

[8] Pamecha, A., Uphoff, I.E., Chirikjian, G.S.: Useful metrics for modular robot motion planning. IEEE Trans. Robot. Automat. 13, 531–545 (1997)

[9] Chirikjian, G.S., Pamecha, A., Uphoff, I.E.: Evaluating efficiency of self-reconfiguration in a class of modular robots. J. Robot. Sys. 13(5), 317–338 (1996)
[10] Kotay, K.D., Rus, D.L.: Scalable parallel algorithm for configuration planning for self-reconfiguring robots. In: Proc. SPIE (Sens Fusion Decentralized Control Robot. Sys. III), vol. 4196, pp. 377–387 (2000)
[11] Nguyen, T., Guibas, L., Yim, M.: Controlled module density, helps reconfiguration planning. In: Proc. Int. Workshop Algorithmic Found Robot., pp. 15–27 (2000)
[12] Butler, Z., Byrnes, S., Rus, D.: Distributed motion planning for modular robots with unit-compressible modules. In: Proc. IEEE/RSJ Int. Conf. Intell. Robot. Sys (IROS), vol. 2, pp. 790–796 (2001)
[13] Stoy, K., Nagpal, R.: Self-repair through scale independent selfreconfiguration. In: Proc. IEEE/RSJ Int. Conf. Intell. Robot. Syst (IROS), pp. 2062–2067 (2004)
[14] Yoshida, E., Murata, S., Kamimura, A., Tomita, K., Kurokawa, H., Kokaji, S.: A Self-Reconfigurable Modular Robot: Reconfiguration Planning and Experiments. Int. J. Robot. Res. 21(10), 903–916 (2002)
[15] Ostergaard, E.H., Tomita, K., Kurokawa, H.: Distributed Metamorphosis of Regular M-TRAN Structures. Distrib. Auton. Robot. Syst. (DARS) 6, 169–178 (2007)
[16] Kurokawa, H., Tomita, K., Kamimura, A., Murata, S.: Toward Flexible and Scalable Self-reconfiguration of M-TRAN. In: Distrib. Auton. Robot. Syst. (DARS), pp. 405–416. Springer, Berlin (2008)
[17] Butler, Z., Kotay, K., Rus, D., Tomita, K.: Generic Decentralized Control for Lattice-Based Self-Reconfigurable Robots. Int. J. Robot. Res. 23(9), 919–937 (2004)
[18] Støy, K.: Using cellular automata and gradients to control self-reconfiguration. Robot. Auton. Syst. 54(2), 135–141 (2006)
[19] Ishiguro, A., Shimizu, M., Kawakatsu, T.: A modular robot that exhibits amoebic locomotion. Robot. Auton. Syst. 54(8), 641–650 (2006)
[20] Inou, N., Fukushima, S., Shimotai, N., Ujihashi, S.: Study of Group Robots Adaptively Forming a Mechanical Structure - Effect of Mechanical Properties of Cellular Robots on Structure Formation. JSME Int. J. C43(1), 127–133 (2000)
[21] Kurokawa, H., Yoshida, E., Kamimura, A., Tomita, K., Murata, S., Kokaji, S.: Self-reconfigurable M-TRAN Structures and Their Walker Generation. Robot. Auton. Syst. 54(2), 142–149 (2006)
[22] Murata, S., Kakomura, K., Kurokawa, H.: Toward a scalable modular robotic system – Navigation, docking, and integration of M-TRAN. IEEE Robot. Automat. Magazine 14(4), 56–63 (2007)
[23] Yim, M., Shirmohammadi, B., Sastra, J., Park, M., Dugan, M., Taylor, C.J.: Towards Robotic Self-reassembly After Explosion. In: Proc. IEEE/RSJ Int. Conf. Intell., pp. 2767–2772 (2007)
[24] Zykov, V., Mytilinaios, E., Desnoyer, M., Lipson, H.: Evolved and Designed Self-Reproducing Modular Robotics. IEEE Trans. Robot. 23(2), 308–319 (2007)

Chapter 8
Self-Organization of Motion

Abstract. In contrast with lattice-type modular robots capable of self-reconfiguration, chain-type modular robots are made for flexible motions, exploiting their multiple degrees of freedom. Motion control is one of the major areas of robotics, and knowledge accumulated through years of experience is applicable to chain type modular robots. Most of such knowledge and methods, however, are suited to a centralized controller based on a complete and precise model. In this chapter, we explore a way to design distributed motion control for modular robots. We start with building a homogeneous distributed control system applicable to any robot configuration. The system then is optimized accordingly to each specific robot configuration.

8.1 Robot Motion Control

We first survey typical designs and methods in robotic motion control. We focus on multi-joint manipulators, legged walking robots, and mobile robots with many degrees of freedom.

8.1.1 Manipulator End Point Control

A robot that corresponds to a human arm is called a *manipulator*. A manipulator consists of several links connected by joints. A serial link type manipulator has its links connected serially, while a parallel link type has its links connected in a loop. Here we focus on serial manipulators.

A typical task of a manipulator is to move its end effector (a device such as a gripper, a paint spray gun, or a welding head) along a designated path. This involves planning of the path (motion trajectory) of the end effector or the whole arm, and then controlling of the joints to realize movement along the path [1].

As a simple example, consider a three joint manipulator restricted to a two dimensional plane (Fig. 8.1). The position of the end effector is geometrically determined by the input angles of the joints, θ_1, θ_2 and θ_3. Namely, we have the following mapping relation:

$$\mathbf{r} = \mathbf{f}(\theta), \quad \theta = (\theta_1, \theta_2, \theta_3), \mathbf{r} = (x, y)^T \tag{8.1}$$

S. Murata and H. Kurokawa: Self-Organizing Robots, STAR 77, pp. 173–209.
springerlink.com

where θ is a set of the three joint angles, and $\boldsymbol{r}$ is a vector representing the position of the end effector with respect to the coordinate frame on the plane. Here, a vector is represented by a column vector[1]. In the case of this manipulator, the mapping $\mathbf{f}$ is given by

$$\mathbf{f}(\theta) = \begin{pmatrix} r_1 \cos\theta_1 + r_2 \cos(\theta_1+\theta_2) + r_3 \cos(\theta_1+\theta_2+\theta_3) + r_{0x} \\ r_1 \sin\theta_1 + r_2 \sin(\theta_1+\theta_2) + r_3 \sin(\theta_1+\theta_2+\theta_3) + r_{0y} \end{pmatrix}, \tag{8.2}$$

where r_i represents the length of each link and $(r_{0x}, r_{0y})^{\mathrm{T}}$ is the position of the fixed joint at the root of the manipulator. The geometric relations like these are called *kinematics*.

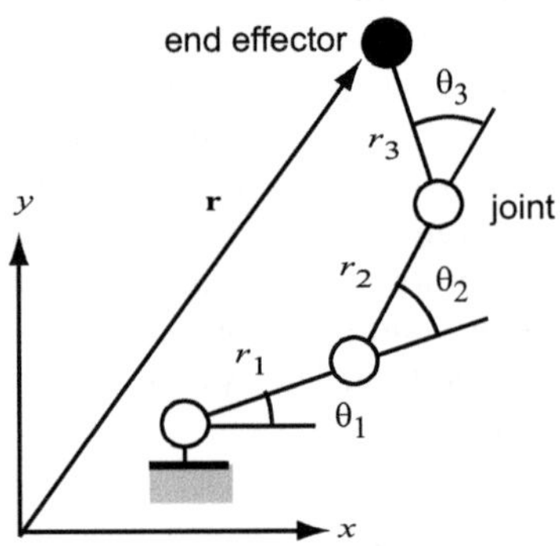

Fig. 8.1 Two dimensional manipulator

In order to maneuver the end effector along a given trajectory, we need to control the joint angles by computing θs that correspond to a given point $\boldsymbol{r}$ on the trajectory. The problem of computing θ from a given $\boldsymbol{r}$ is called *inverse kinematics*, and in general it is not analytically solvable. An alternative method is to slightly displace the target position at a time along the desired motion trajectory and adjust the θs so that the end effector moves toward it. This is achieved by differentiating equations (8.1) and (8.2)

$$d\mathbf{r} = Jd\theta,$$
$$J = (\partial f_i / \partial \theta_j) = \begin{bmatrix} -r_1 \sin\theta_1 - r_2 \sin(\theta_1+\theta_2) - r_3 \sin(\theta_1+\theta_2+\theta_3) & -r_2 \sin(\theta_1+\theta_2) - r_3 \sin(\theta_1+\theta_2+\theta_3) & -r_3 \sin(\theta_1+\theta_2+\theta_3) \\ r_1 \cos\theta_1 + r_2 \cos(\theta_1+\theta_2) + r_3 \cos(\theta_1+\theta_2+\theta_3) & r_2 \cos(\theta_1+\theta_2) + r_3 \cos(\theta_1+\theta_2+\theta_3) & r_3 \cos(\theta_1+\theta_2+\theta_3) \end{bmatrix} \tag{8.3}$$

[1] Mathematically, a set of angles, θ, is not regarded as a vector, whereas its derivative, $\Delta\theta$, is a vector.

where $d\theta$ is a column vector whose elements are $d\theta_i$, and the matrix J is the Jacobian of the mapping **f**. Using the solution[2] $\Delta\theta$ of Eq. (8.3) with small $\Delta\boldsymbol{r}$, the joint angles are adjusted to $\theta+\Delta\theta$. By repeating this computation while controlling the joints, it is possible to realize movement along the trajectory.

The system in this example is a redundant system because the output $\boldsymbol{r}$ has 2 degrees of freedom, whereas the input θ has 3 degrees of freedom. For a redundant system, kinematic equations such as (8.1) and (8.3) do not have unique solutions in general. This means that there are multiple sets of joint angles that give the same end effector position, and that there are multiple sets of joint velocity vectors that give the same end effector velocity. In order to determine a unique solution, extra constraints have to be added. One example of such constraints is that the manipulator should not come in contact with an obstacle. Another example is to require the solution to the linear equations (8.3) to have the minimum norm ($|d\theta|$). We then obtain a unique solution given by

$$d\theta = J^{\#} d\mathbf{r}, \qquad J^{\#} = J^{\mathrm{T}}(JJ^{\mathrm{T}})^{-1}, \tag{8.4}$$

where $J^{\#}$ is the generalized inverse (pseudo inverse) matrix. This is used in many situations.

When a manipulator moves in a three dimensional space and the end effector posture should also be controlled, three more variables for representing the end effector posture have to be added to the three variables representing the end effector position, resulting in more complicated kinematic equations. Still, the control method using the inverse matrix as above is applicable. When the joints are driven by torque actuators and the end effecter needs to follow a trajectory where the position at each time is specified, control should be based not only on kinematics but also on dynamics.

8.1.2 Legged Walking Robots

In this section we consider controlling the walking motion of legged robots. A walking motion changes the position of a robot by maneuvering its multiple legs in a coordinated manner. Each leg of a robot is a multi-joint manipulator, which in principle is maneuvered by kinematic control. In walking motions, each leg alternates between two states: (1) touching the ground and supporting the body (*support leg*) and (2) not touching the ground and swinging to prepare for its next supporting state (*swing leg*). Swing legs contribute little to the motion of the body; it is the support legs that drive the body, but this is not achieved by independent motion of legs [2].

More precisely, walking requires control of the six degrees of freedom of the position and the posture of the body through time. Robots with four or more legs have much larger degrees of freedom, so that the system is largely redundant. Since it is difficult to determine all the redundant degrees of freedom at once, the

---

[2] There can be no solution to the equation (8.3) for certain states of θ. Such a case is called a *singularity* or a *singular point*.

standard procedure is first to determine the timing of switching between swinging and supporting for each leg (which is called a *gait*), then to design the detailed motion of each leg, and finally to determine the motions of the joints.

One of the important things in determining gait is not to let the robot fall. Assume, for simplicity, that the weight of the legs can be ignored compared with that of the body, and the foot, i.e. the portion of each leg in contact with the ground, can be considered as a point. An object, such as a robot, is stable on the ground under gravity on the condition that it contacts with the ground at three or more points and the vertical line from its center of gravity crosses the ground at a point inside the convex polygon made by these points[3]. A gait that always satisfies this condition is called *static walk*. In static walk, the robot will not fall even when it stops.

On the contrary, if the walking motion involves a period when the above stability condition is not satisfied, this is called *dynamic walk*. To realize dynamic walk, motion trajectory planning and stability control considering dynamics is indispensable. Generally speaking, walking with two or three legs must be dynamic walk, and even robots with four or more legs may walk dynamically when walking at a high speed[4].

Fig. 8.2 describes a *crawl gait* as an example of static walk of a four-legged robot. One of the four legs becomes a swing leg in the order shown in Fig. 8.2(a), while the other three legs are used for supporting, in such a way that the center of gravity of the body always stays inside the triangle that they form as in Fig. 8.2(b). Fig. 8.2(c) shows the displacements of the four feet relative to the body. Each foot moves in the same pattern, with each phase a quarter of the cycle out of phase with others. Fig. 8.2(d) shows motion patterns of the joints which move the foot along a rectangular path with respect to the body, when the leg has two degrees of freedom and is capable of planar motion. If the four legs all have this mechanism, the joints of each leg move in the same motion patterns with a certain phase shift.

Fig. 8.3 shows three different types of gaits for six-legged walk. Fig. 8.3(a) shows sequences of gaits which use one swing leg at a time or two simultaneously. Fig. 8.3(b) shows a *tripod gait*, a gait with three legs moving at the same time. This is a simple static walk realized by two groups of three legs making the same triangle with their feet, alternately touching the ground.

As shown by these examples, various gaits are possible even for static walking. Moreover, even after determining the gait, there still are some degrees of freedom in the height and the posture of the body. When each leg has three or more degrees of freedom, there will be extra freedoms in the joint movements.

---

[3] Most control methods of walking robots, either static or dynamic, are based on the idea of ZMP (Zero Moment Point), which is a generalized point where gravity and body's inertial force hit the ground.

[4] Even a robot with two legs is capable of static walk because their feet have non zero areas.

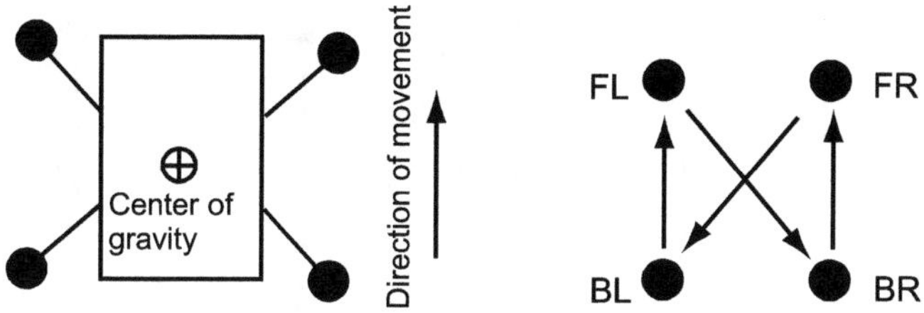

(a) Four-legged robot and its gait

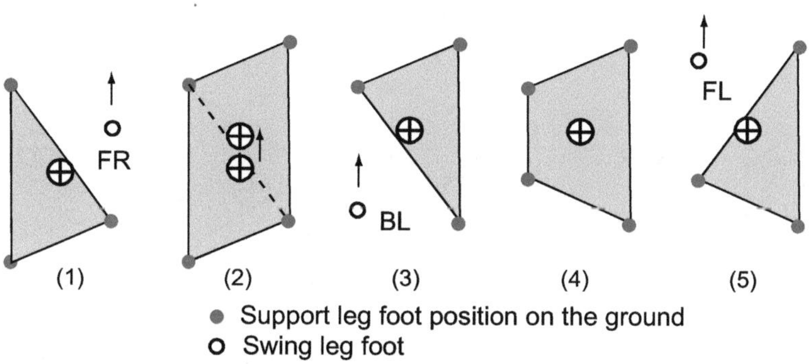

(b) Stable walking pattern: (1) to (5) shows half of a walking cycle, which is followed by the same pattern with left and right reversed. In (2) and (4), the center of gravity shifts while supported by four points.

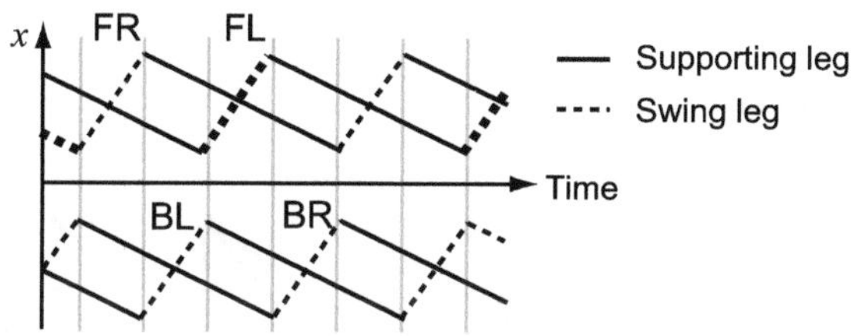

(c) Movements of feet: Variation in time of the forward direction component x of the position of foot relative to the center of gravity.

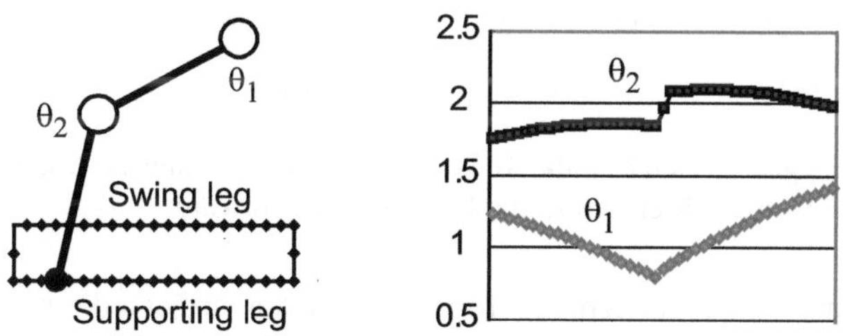

(d) Movements of legs and leg joints: a case of two joint, planar leg

Fig. 8.2 Four-legged walk.

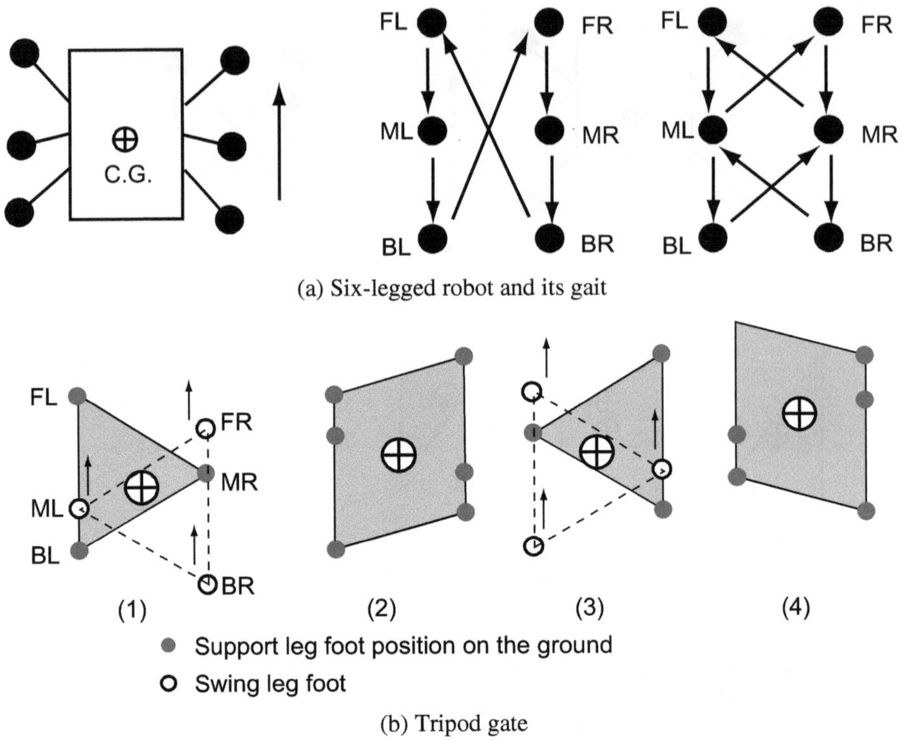

Fig. 8.3 Six-legged walking pattern

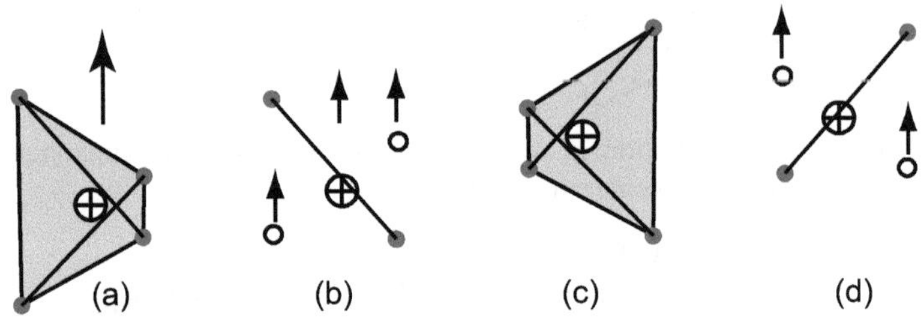

Fig. 8.4 Four-legged dynamic walk. If the robot stops walking at (b), it will fall over backwards, but given a sufficient speed it keeps moving forward.

Fig. 8.4 shows the patterns of foot movements projected on to the ground during dynamic walking of a four-legged robot. The state with two support legs, as in Figs. 8.4(b) and (d), is statically unstable, so the body should sustain a sufficient forward velocity so as to continue forward movement and avoid falling backward, and the swing legs should reach the ground at an appropriate timing so as to avoid falling forward. For dynamic walk, not only the geometric pattern of feet movements should be determined, but also trajectory planning and motion control that take the velocity and the acceleration of the body movements into account are indispensible.

8.1.3 Whole Body Locomotion

In this section we briefly discuss the configurations and the motions of mobile robots other than legged ones that are used occasionally. All examples except the first are statically stable motions.

8.1.3.1 Rolling Motion

A multiple link mechanism forming a loop, whose joint axes are all in parallel as in Fig. 8.5(a), is capable of rolling motion. Assuming it has sufficient width in the direction perpendicular to the page, it does not fall sideways. When the number of joints is sufficiently large, it can be regarded as a smooth ring which is deformed from a perfect circle as in Fig. 8.5(b). In an ideal situation without friction, a perfect circle can keep rolling on the flat ground at constant speed. Friction always exists in reality, and driving force is necessary to compensate for the friction and maintain the movement. The driving force can be generated by keeping the ring's shape as a tilted ellipse as in Fig. 8.5(b). Since the driving force is determined by the ellipticity (the ratio of a and b) and the tilt angle α, feedback control can maintain rolling speed by setting these parameters and controlling each joint angle, i.e., the local curvature, according to the distance from the point touching the ground. [3].

8.1.3.2 Crawler

A structure similar to the closed ring above obtained by deforming the ring into a front half-circle and a rear half-circle connected with straight segments as in Fig. 8.5(c). It can roll and move forward merely by controlling a few joints at the front and the back. This mechanism is called a continuous track or a crawler.

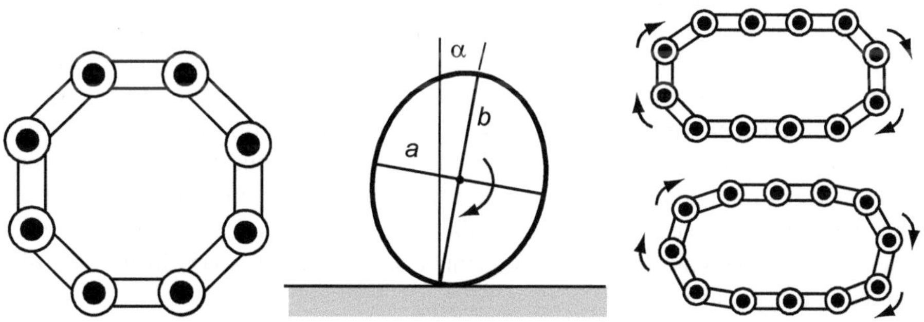

(a) Closed link configuration (b) Elliptic ring (c) Crawler

Fig. 8.5 Rolling motions

8.1.3.3 Traveling Wave on a Serial Link

When a serial chain of multiple links is put on the ground, a vertical traveling wave on the chain makes it move in the same direction as the wave propagation (Fig. 8.6(a)). The wave can be of any form, and the wave in Fig. 8.6(a) made of circular arcs is an example. This movement does not cause slipping at points touching the ground, and locally at each of these points, the motion is equivalent to the rolling motion of a loop.

Representing each point by the distance s measured along the body from one end, the function $\theta(s, t)$ representing the joint motion (the wave curvature) is given as

$$\theta(s, t) = f(t - Ts/\lambda) , \tag{8.5}$$

where T is the period of the periodic function f, and λ is the length of the body segment that corresponds to one wave cycle. The same periodic function f is used for every joint with a phase offset proportional to the distance. The wave in Fig. 8.6(a) is generated by the function in Fig. 8.6(b). More practical functions suitable for control are trapezoidal, triangular and sinusoidal functions. With a sinusoidal function of joints, the wave shape of the body becomes almost sinusoidal.

Note that, with the above waves, the head and the tail of the robot do not stay on the ground but move up and down, which serves no actual purpose. Fig.8.6(c) shows a wave pattern that is modified so that the head and the tail stay on the ground. This wave pattern is obtained by the concatenation of (1) growing a wave with increasing wave length and amplitude at the tail, (2) sending it towards the head, and (3) the reverse process of (1) at the head, in which the wave attenuates and disappears.

8.1.3.4 Motions of Snakes and Fish

In the motion described in the previous section, the wave occurs in a plane perpendicular to the ground, and the robot moves forward due to the rolling motion without slipping at the contact points with the ground. A snake type motion is realized by a traveling wave of similar shape, but which moves in the plane parallel to the ground [4]. In this case, the entire surface of one side of the robot is in contact with the ground. When the ground friction in the longitudinal direction of the body is less than that in the lateral direction, the result is slipping motion along the body while the whole wave-shaped body moves in the opposite direction of the wave propagation. A fish whose body is flat and long receives friction from the water in a similar manner as this; it can swim in the opposite direction to the wave.

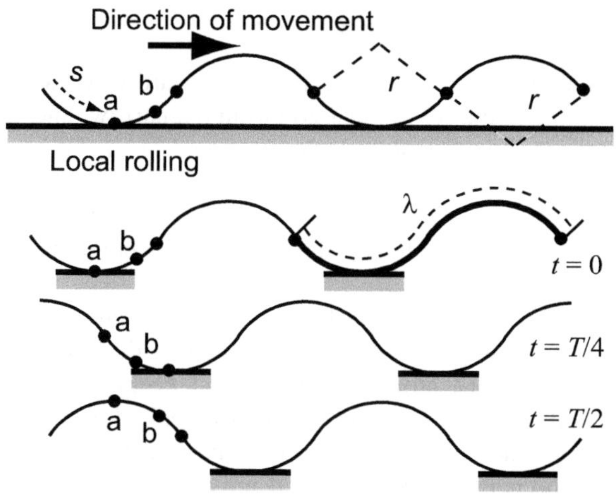

(a) Traveling waves of circular arcs: wave form and propagation

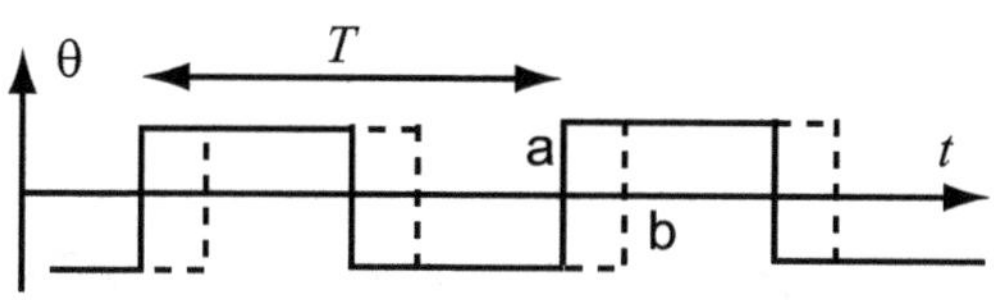

(b) Change in joint angles: the changes in joint angles at the points a and b in (a)

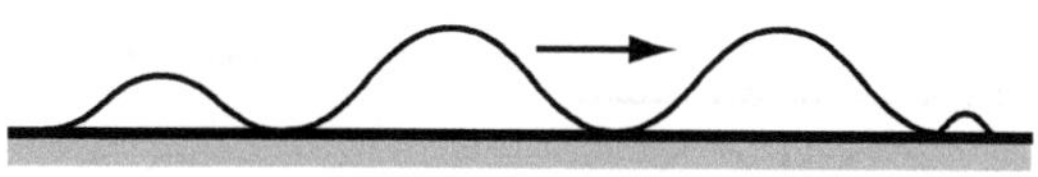

(c) Superposition of solitary waves: modified wave form

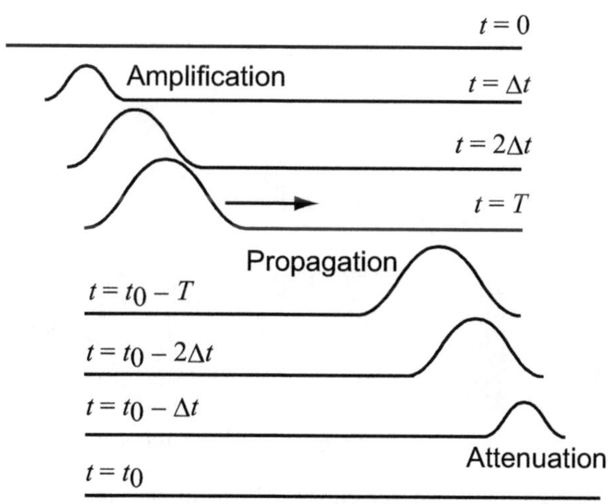

(d) Transmission of solitary wave

Fig. 8.6 Motions of traveling waves

8.1.3.5 Combinations of Traveling Waves

A motion that combines a forward wave and a backward wave is also possible. Let us consider the case that the wave must go through a narrow hole as in Fig. 8.7(a). In the motion described in Fig. 8.6(c), the front end of the body stays in contact with the ground, but obviously the body cannot move into the hole. The wave has to go through the hole at its minimum height, so that this point has to move in the direction opposite to that of the body. Moreover, at this point the body needs to be slipping.

It is possible to generate a backward wave at the hole and forward waves at the rest of the body by dividing the robot in two, keeping the division point between them progressing backward along the robot, and keeping waves in both parts progressing forward (Fig. 8.7(a)). This can be also applied for the case where there is a step instead of a hole; by extending the division point into a segment of a shape that covers the height of the step and moving this segment backward, it is possible to climb up the step, keeping the whole body moving forward (Fig. 8.7(b)). Fig. 8.7(c) shows an experiment in which a robot consisting of ten M-TRAN modules climbed up a step in this way.

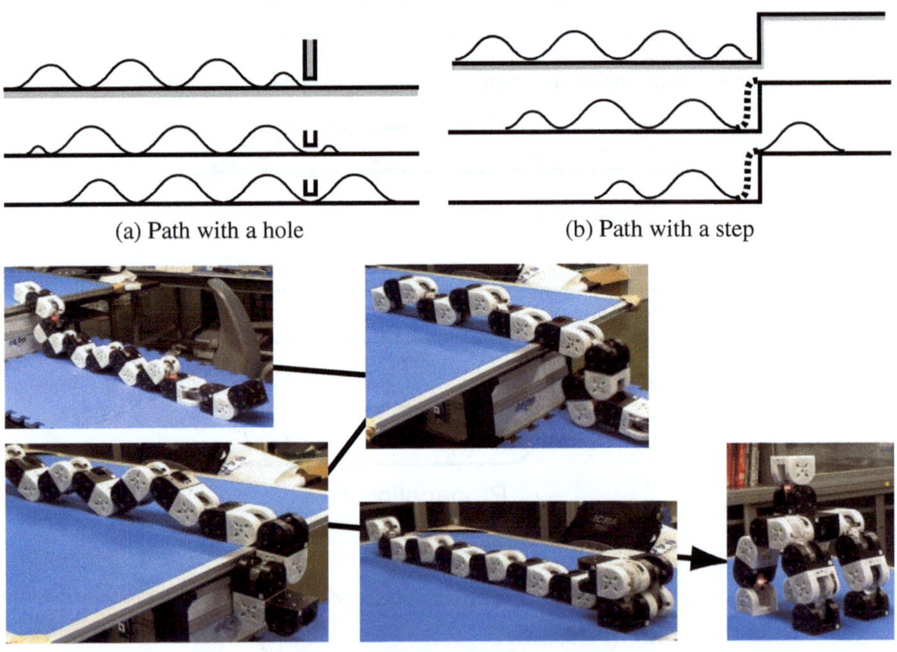

(a) Path with a hole

(b) Path with a step

(c) Step-climbing of M-TRAN robot

Fig. 8.7 Passing through a hole and step climbing by partitioned traveling waves

8.1.4 Design of Motion Control Systems

As we have seen so far, there are various issues to be considered in designing robotic motion control; kinematics for manipulators, gait patterns for legged robots, whole body motions, and so on. However, all of them are designed assuming a centralized control as illustrated in Fig. 8.8(a). Our next question is if it is possible to realize this control using a distributed control as in Fig. 8.8(b).

For the kinematic control of manipulators, distributed control is not suitable, as long as it is strictly based on (8.1) and (8.3). This is due to the fact that each element of the Jacobian matrix in (8.3) is given as a function of multiple joint angles and therefore it is not possible to decouple them and solve linear equations for each joint[5]. This also applies to the case of legged robots whose gaits and joints are designed and controlled based on precise kinematics.

Then what about manipulators with a larger number of joints or mobile robots with many legs like centipedes? Is it appropriate to use such centralized, precise control for these cases, too? In the case of manipulators with numerous joints, solving inverse kinematics may not be always the most appropriate method of control, because of increase in computational complexity with increase in the number of joints. Also, for the case of multi-legged robots, if the goal is simply to move forward, it may not be necessary to compute the movements of each joint in advance and maneuver them precisely.

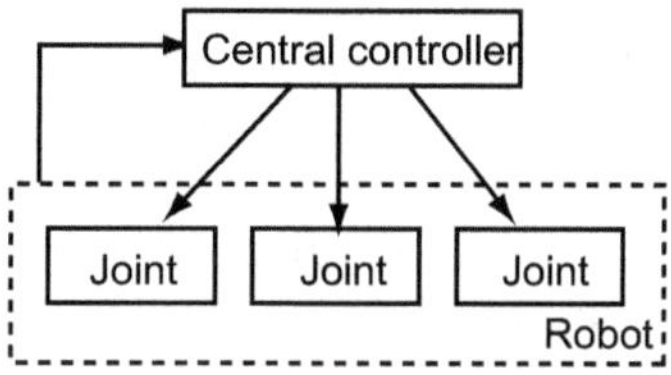

(a) Centralized control

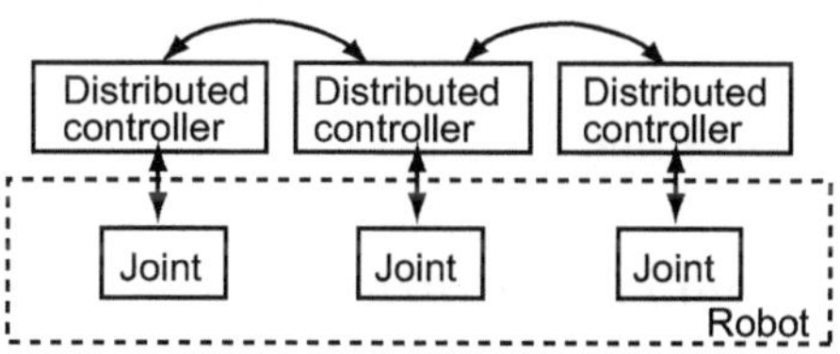

(b) Distributed control

Fig. 8.8 Hierarchy and distribution of control

In fact, the problem that must be addressed is not whether the control system structure should be centralized or distributed, but a methodology issue: what kind of movement should be realized and how to realize it. All of the cases we

[5] There have been attempts of distributed inverse-kinematic control using iteration [5].

discussed so far are examples of determining joint movements based on precise motion models. In order to achieve this precision, constraints to the equations are required depending on the number of redundant degrees of freedom, which increase in proportion to the number of joints. In the case of four-legged robots, the crawl gait in Fig. 8.2 seems a natural choice. In the case of six-legged robots, however, there are many possible gaits so that we need appropriate criteria, or equivalently, constraints for a unique choice. When there are even more legs, more constraints have to be given. Design by distributed control and self-organization which we discuss in the next section is advantageous for such cases.

8.1.5 Distributed Motion Control of Modular Robots

Various chain-type modular robots we introduced in Chapter 6 can realize robotic motions with various configurations. M-TRAN modules also work as a chain-type modular robot. Those systems are built as distributed systems consisting of microcomputers mounted in all the modules and communication means connecting the modules. In most of the studies, however, robotic motions described above such as multi-legged walking, traveling waves, and crawlers were used, which were in effect equivalent to centralized control. Even with distributed modules, all the modules need to globally synchronize and to follow a specific (global) model.

Centralized control based on a precise model is certainly useful. However, a chain-type modular robot has many possible configurations, and once the configuration is changed, the motion model has to be selected and the motion pattern of joints has to be designed again from scratch. Moreover, there is no guarantee that a model suitable for the new configuration can be found easily. In order to bring out the potential of chain-type modular robots, a common framework is desired from which a controller can be designed for any configuration. The method we explain in the next three sections is an attempt to create such a framework.

8.2 Coupled Oscillators

8.2.1 Synchronization by Diffusion

As we discussed in Chapter 4, diffusion is a general principle for making the state of a system homogeneous. This principle can be applied also to synchronization, because synchronization is a process making oscillating objects have the same frequency and phase [6-8].

Diffusion interaction of discrete components is described again as follows:

$$\mathrm{d}s_i/\mathrm{d}t = D\,\Sigma_j(s_j - s_i) + f_i\,, \tag{8.6}$$

where s_i is a state variable of the ith component, D is the diffusion coefficient, and the sum in the first term on the right hand side is taken over all the components connecting to the ith component.

As an example of synchronization, consider a situation where two components generate and exchange periodic pulses. Each of the components can adjust their pulses as follows: on receiving a pulse, if there is still some time scheduled until the next pulse that the component sends, then it sends the next pulse earlier; whereas if its next pulse is about to be sent but a pulse has not been received yet, then it sends the next pulse a little later. In this way, the time delay of pulses between the two components can be reduced. This is none other than a diffusion process. Let us look at this mechanism in more detail.

Consider an oscillator (a *phase oscillator*) whose state is given by a phase angle. Letting the phase angle ϕ be the state variable and ω be the angular frequency, the simplest equation for the oscillation is given as:

$$\mathrm{d}\phi/\mathrm{d}t = \omega. \tag{8.7}$$

Now, connect two oscillators using diffusion as follows:

$$\mathrm{d}\phi_i/\mathrm{d}t = \omega_i + D\,(\phi_j - \phi_i)\,, \quad i=1,2 \quad j=2,1, \tag{8.8}$$

where the two oscillators are distinguished by indices 1 and 2, and the difference of phase angles is computed modulo 2π so that the resulting value is in the range $[-\pi, \pi]$. This modulo calculation is one of the differences between this system and a usual diffusion system. In addition, the phase of the oscillator, ϕ_i, keeps increasing and the sum of the left hand sides of the equations for the two oscillators is not equal to zero. Taking the sum and the difference of the two systems, we obtain

$$\mathrm{d}(\phi_2+\phi_1)/\mathrm{d}t = (\omega_2+\omega_1)\,,$$

$$\mathrm{d}(\phi_2-\phi_1)/\mathrm{d}t = (\omega_2-\omega_1) - 2D\,(\phi_2-\phi_1)\,. \tag{8.9}$$

From this we see that the two oscillators converge to steady oscillation at the average frequency of $(\omega_2 + \omega_1)/2$ while keeping the phase difference of $(\omega_2 - \omega_1) / (2D)$.

In the case when there are many oscillators, if we let ω_0 be the average frequency, and substitute $\phi_i = \omega_0 t + \alpha_i$ in equation (8.8), we obtain

$$\mathrm{d}\alpha_i/\mathrm{d}t = (\omega_i - \omega_0) + D\,\Sigma_j\,(\alpha_j - \alpha_i)\,. \tag{8.10}$$

This shows that when many phase oscillators are connected together, they all end up in a synchronized state oscillating at the same frequency while keeping a constant phase offset. In particular, when the frequencies of all oscillators are the same, this leads to synchronization with no phase offset over the entire system.

It is to be noted that the validity of the above analysis depends on the connection topology and initial states of the oscillators. The behavior described by

(8.10) may not converge to a state with zero phase difference even when all the components' frequencies are the same. For example, consider the five components connected in a loop as in Fig. 8.9(a), supposing that they have the same frequency ω and diffusion coefficient D is 1. Now, by representing the phase angle ϕ by a point on the unit circle, the diffusion terms acting at node 2, for instance, can be expressed as arrows in Fig. 8.9(b). Each arrow is of the length of the arc between points, and hence the diffusion effect works as if these arrows are springs.

All the phase points in this system converge to a single point if its initial configuration is like that of Fig. 8.9(b). From other initial configurations, however, the system may end up in a stable equilibrium state like that of Fig. 8.9(c), in which the spring-like effects between each other balance out.

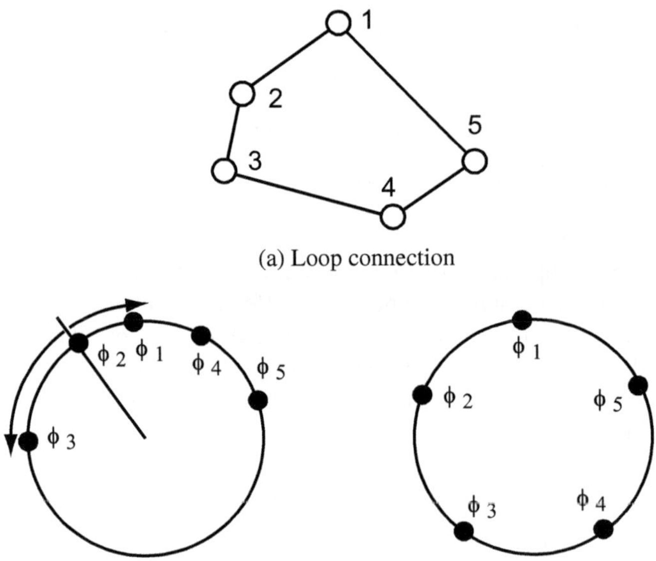

(a) Loop connection

(b) Diffusion interaction (c) Circular equidistant configuration

Fig. 8.9 Problem by loop connection

In this way, simple diffusion interactions of phase angles do not assure convergence to a single point. This problem is alleviated by changing the spring behavior of the diffusion term from a standard straight line (Fig. 8.10(a)) to a broken line f (Fig. 8.10(b)). The system then is described as

$$d\phi_i/dt = \omega_i + D\,\Sigma_j f(\phi_j - \phi_i). \tag{8.11}$$

By using this function, the circular equidistant equilibrium state in Fig. 8.9(c) becomes unstable and the system converges to a single point due to the nonlinearity of f that now has negative slopes. When the number of nodes on the circle is at most n, they are guaranteed to converge to a state with zero phase difference from any initial state if $\varphi < \pi / n$ [9].

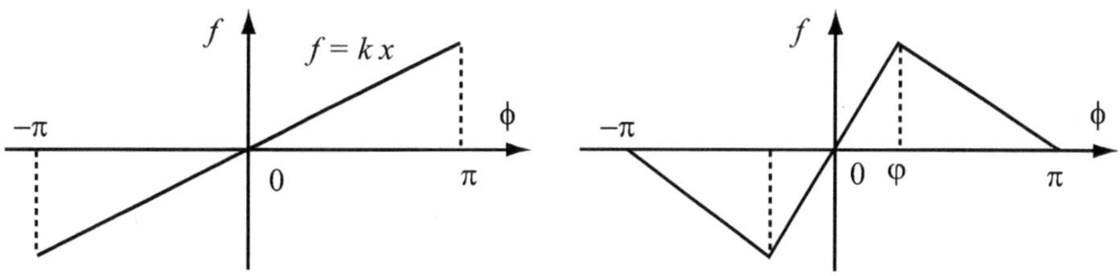

(a) Linear (b) Nonlinear (piecewise linear)

Fig. 8.10 Non-linear interaction

8.2.2 *Entrainment*

Multi-joint coordination of a robot in Sec. 8.1 may be realized by synchronized oscillators such that each joint is controlled by an appropriate function such as Eq. (8.5) with appropriate phase offset. This, however, is a simple feed-forward (open-loop) control. For feedback control, phase information of actual periodic motions needs to be fed back to the oscillators. For example, if a leg of a robot has a sensor for detecting the ground contact, this on/off signal can be used as a feedback signal because it carries explicit phase information. However, signals usually available in a physical system such as joint angles and forces change periodically, and obtaining phase angles from those signals is not straightforward.

In fact, there is a method to execute synchronization and feedback at the same time using a phenomenon called *entrainment* of oscillators, which is realized by making oscillating variables, such as joint angles and forces, interact directly with oscillators. We here give a simple explanation of this phenomenon. In the rest of this section, we assume that all oscillators have the same frequency.

We use an oscillator described by a second-order differential equation instead of the phase oscillator. Given a second-order differential equation

$$\mathrm{d}^2x/\mathrm{d}t^2 = F(x, \mathrm{d}x/\mathrm{d}t)\,, \tag{8.12}$$

and by introducing two variables defined as

$$x_1 = x\,,$$

$$x_2 = \mathrm{d}x/\mathrm{d}t\,,$$

(8.12) can be converted into first-order differential equations:

$$\mathrm{d}x_1/\mathrm{d}t = x_2\,,$$

$$\mathrm{d}x_2/\mathrm{d}t = F(x_1, x_2)\,. \tag{8.13}$$

We start with the simplest sinusoidal oscillator (harmonic oscillator)

$$d^2x/dt^2 = -\omega^2 x\,, \tag{8.14}$$

where ω is the angular frequency. Similarly to the above, this second-order equation can be reformulated as first-order differential equations:

$$dx_1/dt = -\omega x_2\,,$$

$$dx_2/dt = \omega x_1\,. \tag{8.15}$$

Putting the two variables together as a column vector, we have

$$d\mathbf{x}/dt = \omega|\mathbf{x}|\,\mathbf{n}, \tag{8.16}$$

where $\mathbf{x} = (x_1, x_2)$, $|\mathbf{x}| = \sqrt{x_1^2 + x_2^2}$ and $\mathbf{n} = (-x_2, x_1)/|\mathbf{x}|$.

The space of such vector variables is called a *phase space*, and a two dimensional phase space is called a *phase plane*. In the phase plane, the vector $\mathbf{x}$ is represented as an arrow from the origin (0, 0) pointing at (x_1, x_2) as in Fig. 8.11(a). The vector $\mathbf{n}$ is a vector of length 1 in the direction obtained by rotating $\mathbf{x}$ to the left by 90 degrees. Thus the vector $d\mathbf{x}/dt$ is orthogonal to $\mathbf{x}$ in the phase plane, and the trajectory of $\mathbf{x}$ (called *solution trajectory*) is a circle centered at the origin, whose radius is determined by an initial condition.

This differential equation can be expressed in polar coordinates (r, ϕ) as

$$d\phi/dt = \omega, \quad dr/dt = 0\,. \tag{8.17}$$

Considering ϕ only, this is the same as (8.7).

Now, we add the following non-linear term in order to get a unique trajectory:

$$d\mathbf{x}/dt = \omega|\mathbf{x}|\,\mathbf{n} + \mathbf{g}\,, \tag{8.18}$$

where

$$\mathbf{g} = g(|\mathbf{x}| - a)\mathbf{x}/|\mathbf{x}|\,,$$

$$g(r)\begin{cases} <0, & r>0 \\ =0, & r=0 \\ >0, & r<0 \end{cases}. \tag{8.19}$$

The vector $\mathbf{g}$ is in the direction of the moving radius, which makes the system's trajectory converge to a circle with radius a (Fig. 8.11(a)). The closed trajectory like this circle to which other trajectories converge is called a *limit cycle*. In polar coordinates, the system differs from the original one in (8.17) only in the second equation for r:

$$dr/dt = g(r - a)\,. \tag{8.20}$$

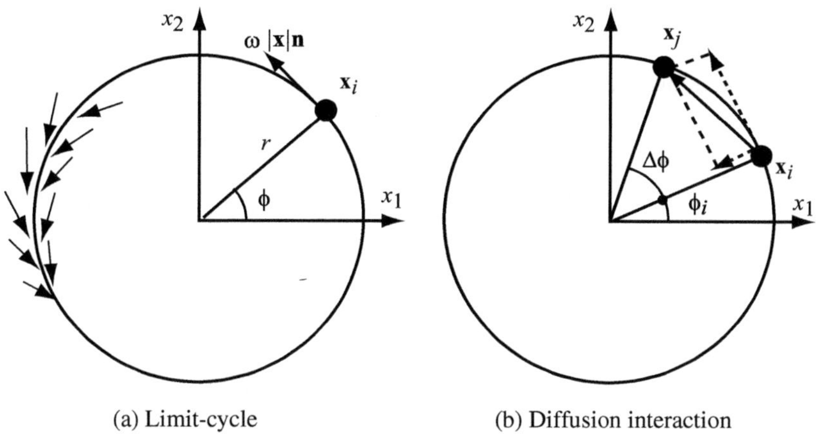

Fig. 8.11 Entrainment of oscillators. (a) Limit-cycle (b) Diffusion interaction

Now, we let two oscillators interact directly through the state variable $\mathbf{x}$

$$d\mathbf{x}_i/dt = \omega|\mathbf{x}_i| \, \mathbf{n}_i + \mathbf{g} + k(\mathbf{x}_j - \mathbf{x}_i) \,, \tag{8.21}$$

where the two oscillators are distinguished by the indices i and j. The third term on the right represents the diffusion interaction. Fig. 8.11(b) shows the vector of the diffusion term in the phase plane. The interaction term can be decomposed into components along the radius and the tangent as drawn in the figure. The component along the radius always works in the direction toward the origin, and in situations where the restoring effect of $\mathbf{g}$ is large or the two states are close to each other, the radius r changes only a little.

The component along the tangent is obtained by calculating the inner product with the vector $\mathbf{n}_i$:

$$d\mathbf{x}_i/dt \cdot \mathbf{n}_i = \omega|\mathbf{x}_i| + k \, |\mathbf{x}_j - \mathbf{x}_i| \cdot \mathbf{n}_i. \tag{8.22}$$

Setting r to be constant, this can be rewritten in terms of ϕ as

$$d\phi_i/dt = \omega + k \sin(\phi_j - \phi_i). \tag{8.23}$$

This equation can be regarded as (8.11) where the piece-wise linear function f is smoothed out with the changeover point set to be $\pi/2$. When the phase difference is sufficiently small, the system behaves like a linear diffusion process, which means that the two oscillators synchronize and both of their radius r become a.

Now let us change the form of diffusion interaction to be more realistic. Removing $\mathbf{x}_i$ from the third term of (8.21), we obtain

$$\mathrm{d}\mathbf{x}_i/\mathrm{d}t = \omega|\mathbf{x}_i/\ \mathbf{n}_i + \mathbf{g} + k\,\mathbf{x}_j, \tag{8.24}$$

which can be transformed to

$$\mathrm{d}\mathbf{x}_i\,/\mathrm{d}t = \omega|\mathbf{x}_i/\ \mathbf{n}_i + (\mathbf{g} + k\,\mathbf{x}_i) + k(\mathbf{x}_j - \mathbf{x}_i)\,. \tag{8.25}$$

With this, the radius of the limit cycle may change because of the deviation in the second term on the right from (8.21), but the phase angles ϕ_i are synchronized as above in accordance with (8.23).

Next, we remove the component x_{j2} from $\mathbf{x}_\mathrm{j} = (x_{j1}, x_{j2})$ in the third term of equation (8.24) to obtain

$$\mathrm{d}\mathbf{x}_i\,/\mathrm{d}t = \omega|\mathbf{x}_i/\ \mathbf{n}_i + \mathbf{g} + k\,(x_{j1}, 0)\,. \tag{8.26}$$

The third term on the right works only horizontally in the phase plane, which is not always an interaction that causes $\mathbf{x}_i$ to approach $\mathbf{x}_j$, as shown in Fig. 8.12. Synchronization cannot be confirmed merely by observing each oscillator on the phase plane. If we assume that $\mathbf{x}_i$ and $\mathbf{x}_j$ are on the same circle, the component along the tangent line of Eq. (8.26) given as

$$\mathrm{d}\mathbf{x}_i\,/\mathrm{d}t \cdot \mathbf{n}_i = \omega|\mathbf{x}_i| + k(x_{j1}, 0) \cdot \mathbf{n}_i \tag{8.27}$$

can be rewritten by representing x_{j1} and $\mathbf{n}_i$ in terms of the phase angles:

$$\mathrm{d}\phi_i\,/\mathrm{d}t = \omega - k\,\sin\phi_j\,\cos\phi_i.$$

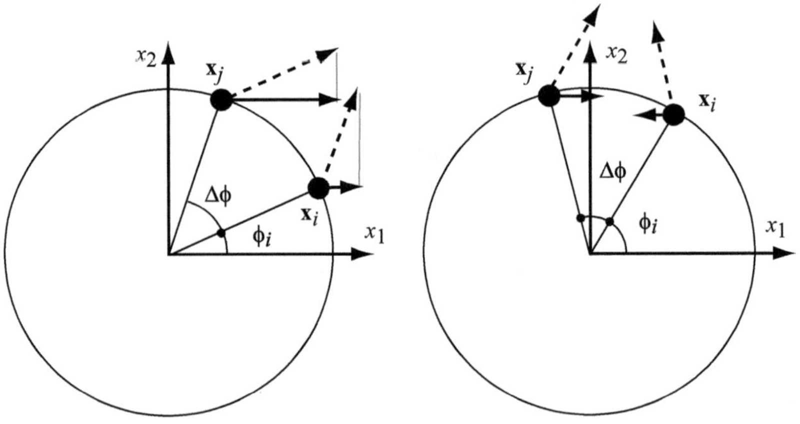

Fig. 8.12 Entrainment of oscillators by modified interaction

Using this, the difference of two oscillators is given as follows:

$$\begin{aligned} d(\phi_i - \phi_j)/dt &= -k\,(\sin\phi_j \cos\phi_i - \sin\phi_i \cos\phi_j) \\ &= k \sin(\phi_j - \phi_i)\,, \end{aligned} \tag{8.28}$$

which means that the phase difference converges to zero due to an effect similar to the previous example, and that the two oscillators eventually synchronize.

We have explained qualitatively why coupled oscillators synchronize. Such synchronization is an example of a common phenomenon in nonlinear dynamics called *entrainment*. If we apply the same interaction to a more general oscillator, we will have

$$\begin{aligned} dx_{i1}/dt &= F(x_{i1}, x_{i2}) + \Sigma_j\, k_{ij}\, x_{j1} \\ dx_{i2}/dt &= G(x_{i1}, x_{i2}). \end{aligned} \tag{8.29}$$

In this case, the limit cycle of each oscillator may not be circular, and there may be more than one limit cycle. With the interaction, synchronization may or may not occur, or even if synchronization is achieved, the limit cycle of each oscillator may differ. Moreover, cycle doubling, intermittent transition from one limit-cycle to another, or non-periodic chaotic behavior may be exhibited. Such complex and interesting phenomena are the central themes of non-linear dynamical systems, especially the theory of bifurcation and chaos[6].

What is important in design and control of mechanical systems is that oscillators with similar frequencies tend to synchronize through interaction of a single output x, and that this can be utilized for motion control of a robot.

8.2.3 How to Introduce Phase Offsets

So far we have been discussing situations where the phases are all to be the same. However, in order to use coupled oscillators for motion control of a robot, each oscillator, corresponding to each joint, needs to keep an appropriate phase offset with others. There are several possible ways to generate phase offsets.

1. Specify phase offset directly for each component. In the case of a phase oscillator, it is easy to set both phase offset and the oscillation (amplitude) function arbitrarily as in (8.5). When the system is described by two variables as in (8.29), an arbitrary phase offset can be introduced by regarding a linear combination of two variables, $a_1 x_1 + a_2 x_2$, as an output with appropriate coefficients a_1, a_2.
2. Introduce phase gradient by assigning a different frequency to each oscillator, as demonstrated by (8.10).

---

[6] For further reading refer to [10-12].

3. Use the circular equidistant equilibrium state shown in Fig. 8.9(c), which we pointed out as a problem in synchronization. The same diffusion system may converge to either a state of synchronization or the circular equidistant state, depending on the initial state. Therefore, to utilize this convergence phenomenon, some additional mechanism is required to achieve the desired equilibrium state.
4. Set the interaction coefficient k to be negative so that interacting oscillators move in opposite directions in the phase space. Fig. 8.13(a) shows that, in the case where two oscillators are coupled with negative diffusion, they come to be in anti-phase. It is easier to understand to regard this as each oscillator is pulled by the other's anti-phase shadow shown by white circles in Fig. 8.13 (a), rather than as each repels the other. Fig. 8.13(b) shows three oscillators coupled together by negative coefficients. They converge to be 120° out of phase with each other. Note that there are two possible ways to arrange them on the circle. Fig. 8.13(c) is an example of how the coefficient k influences the phase difference.
5. Synchronization can be regarded as a diffusion process of one variable. As we explained in Chapter 4, a two variable reaction diffusion system may exhibit a Turing pattern, an oscillating pattern, or a propagating wave. Eqs. (8.21) and (8.29) are indeed reaction diffusion equations, so a propagating wave can be generated by an appropriate reaction function.

Among the above, the second and the fourth methods are currently more realistic, considering their system homogeneity and the stability of their equilibrium state. In the design method we explain in Section 8.5, we adopt the fourth method[7].

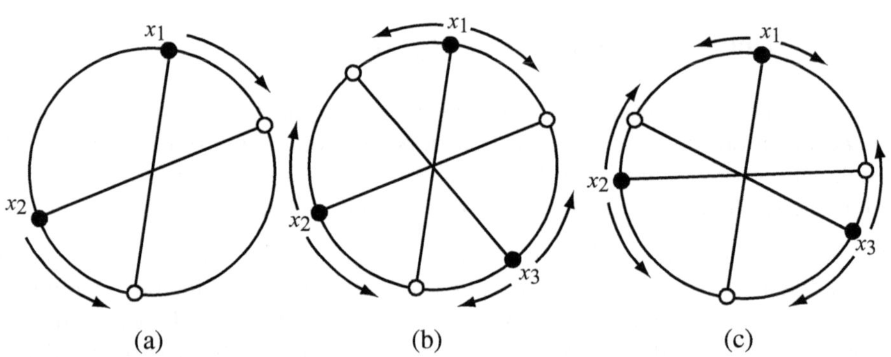

Fig. 8.13 Setting phase difference using negative diffusion coefficient. (a) Two oscillators are coupled by the same negative coefficients. (b) Three oscillators are symmetrically coupled with the same negative coefficients. There is another steady state where three states are in reverse order on the circle. (c) The coupling coefficients of (b) are made asymmetric.

[7] The controller for SlimeBot in Section 6.4.15 uses the second method [13].

8.3 Motion Control Using Coupled Oscillators

There are mechanisms of motion generation in many biological organisms which utilize biological oscillators connected to biological motor systems. Biological oscillators are made of several neurons and are called *Central Pattern Generators (CPGs)*. Many studies have been done on details of neural oscillators and nerve electric potential oscillations in various biological systems [14]. The word "central" in its name does not indicate the existence of a single oscillator in charge of central control, but rather indicates that the network of neurons plays a central role in generating oscillations suitable for the motion.

There has been research on methods for controlling robot motions based on CPG, which we call "CPG control" in this book [15, 16]. In a CPG control, a system of coupled oscillators as described in the previous section is constructed, and using those oscillators the set of joints of a robot is controlled. Various kinds of motions can be realized by maintaining the phase differences among the oscillators appropriately. If coupled oscillators' outputs are merely connected to joints and feedback is not given to the oscillators, coordinated motion of the entire system may not be attainable. Connecting oscillators and joints in both directions is often necessary.

8.3.1 Connection with Physical Systems

Consider a system consisting of an oscillator and a physical pendulum connected together. The left hand side of the system in Fig. 8.14 is a pendulum driven by a torque motor and the right hand side is the oscillator circuit. If there is no connection from the pendulum to the oscillator, then this is a conventional forced oscillation problem. Thus, the free oscillation of the pendulum converges as time passes to follow the oscillator's output with a particular amplitude and a particular phase offset determined by factors such as the damping factor and the difference in frequency of the two systems.

In the case of two-way connection shown in Fig. 8.14, the state variable x_1 of the oscillator works as the driving force of the pendulum, while the swing angle θ acts upon the oscillator. If the two systems are expressed in equations of the same form

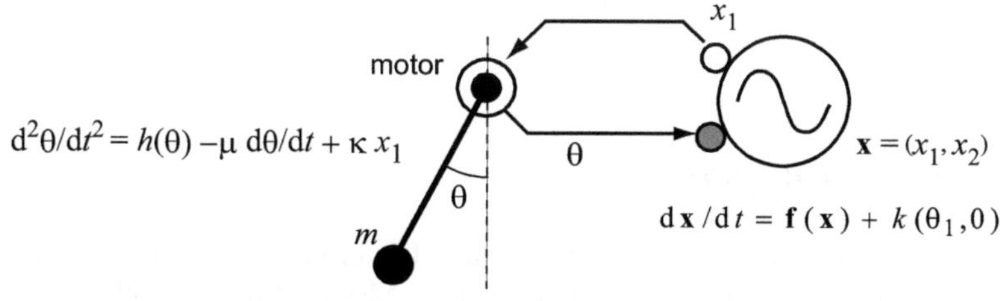

Fig. 8.14 Interaction between oscillator and pendulum

$$d\mathbf{x}/dt = \mathbf{f}(\mathbf{x}) + k(\theta_1, 0) \quad (8.30)$$

and

$$d\theta/dt = \mathbf{f}(\theta) + k(x_1, 0) \quad (8.31)$$

their behavior is also the same whether the oscillator is built as an electrical circuit, a computer program, or a physical pendulum, and therefore just by connecting the two systems, they synchronize. However, in the case of the physical pendulum in Fig. 8.14, the input to the motor is a torque, and its motion equation is given as

$$d\theta/dt = \mathbf{f}'(\theta) + k(0, x_1) \quad (8.32)$$

which has a form different from that of (8.31): the interaction is given to the second term. So, for now, we rewrite (8.32) using the variable substitution $\phi = (\phi_1, \phi_2) = (-\theta_2, \theta_1)$ as follows:

$$d\phi/dt = \mathbf{g}(\phi) + k(x_1, 0) \quad (8.33)$$

The new variable ϕ represents the state vector 90 degrees ahead of θ. The interaction between equation (8.30) and (8.33) means that x is pulled by θ and ϕ is pulled by x. Fig. 8.15 shows the balanced state reached when the frequency of x is larger than that of θ. It can be seen that they synchronize at a frequency between their original frequencies while keeping a certain phase difference.

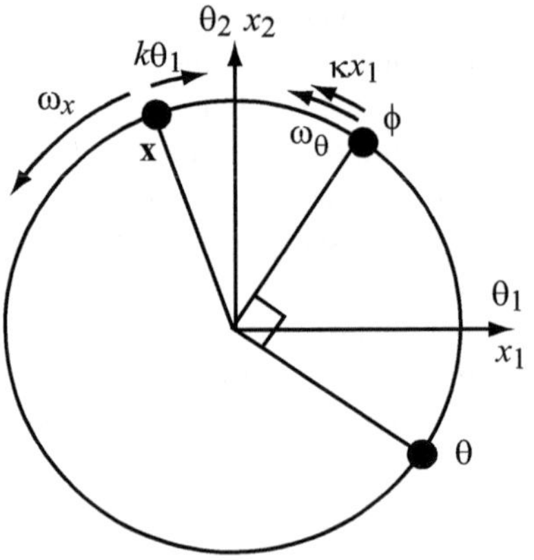

Fig. 8.15 Synchronization of a physical system

8.3.2 *Global Entrainment*

The above discussion explains only the bilateral interaction between an oscillator and a physical system of single degree of freedom. When robotic motion is generated by several oscillators and joints, the periodic motion of the entire robot also affects every joint and oscillator.

For example, in the case of the locomotion of a multi-legged robot, due to the walking motion in which the swing leg alternates with support legs one after another, the body of the robot swings back and forth and side to side like a pendulum. This means that the entire robot behaves like an oscillator as in Fig. 8.16. This oscillation of the entire body reaches the oscillator circuits through the controllers in each leg and affects all the oscillator circuits, which results in entrainment. This entrainment affects both the phase and the amplitude of joint motions because, from the perspective of each joint, motions of the entire robot appear as changes in the load and also changes in the inertia m in Fig. 8.14. Entrainment caused by the motions of the entire system like this is called *global entrainment*, as opposed to *local entrainment* caused by oscillator coupling, which we explained earlier[8].

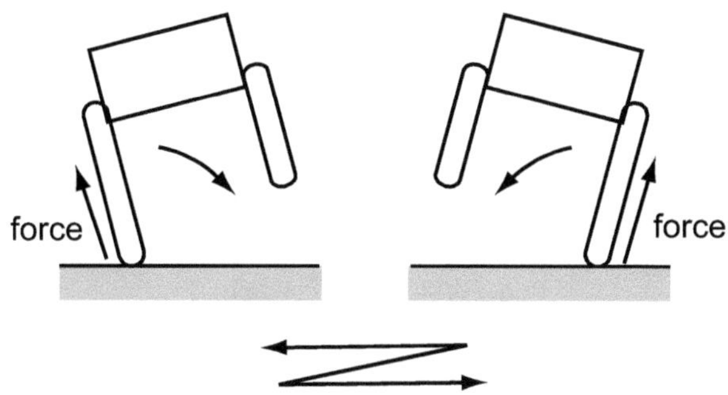

Fig. 8.16 Global entrainment

8.3.3 Neural Oscillator

In this section, we introduce *Matsuoka's oscillator* [18] as one of the mathematical models of biological CPG. It consists of two subsystems connected in anti-phase. A muscle, which is a biological actuator, can exert force only by contraction, and in order to move a joint in two opposite directions, two types of muscles, a flexor and an extensor, need to be controlled individually by different neurons. In order to simulate such a system, two subsystems are connected in anti-phase. Also, non-linear characteristics are introduced to simulate the characteristics of neuron connection. Nevertheless, this system can be understood qualitatively using linear approximation.

[8] Global entrainment based on bidirectional connection between an oscillator and a joint motor is not always possible when motors are controlled by local servo controllers. Even in such a case, some means of global feedback is necessary for the CPG control to work properly, which uses information on the whole body and environment. An example is the use of reflex control, where external information such as touch sensing at the foot causes phase of the leg motion shift or reset [17]. A CPG control is not only a way to maintain mutual synchronization but also a way to maintain whole body motion.

Matsuoka's oscillator is defined by the following differential equations:

$$
\begin{aligned}
&\tau\,\mathrm{d}u_i/\mathrm{d}t = -u_i - \beta v_i + u_{\mathrm{e}} - \alpha_i \tilde{y}_i + a s_i + k_i f_i\,,\\
&\tau'\mathrm{d}v_i/\mathrm{d}t = -v_i - y_i\,,\\
&y_i = \begin{cases} 0, & u_i \le 0 \\ u_i, & u_i > 0 \end{cases},\\
&\tilde{y}_i = \begin{cases} y_2, & i = 1 \\ y_1, & i = 2 \end{cases},\\
&i = 1, 2,\\
&k_1 > 0,\ k_2 = -k_1\,.
\end{aligned}
\tag{8.34}
$$

Here, the first two equations for each i, represent two subsystems that have exactly the same characteristics. The pair (u_i, v_i) is the state variable of each subsystem, y_i and $\tilde{y}_i$ are dependent variables defined in the third and fourth equations, and s_i and f_i are inputs that we define later. The other coefficients τ, τ', β, u_{e}, α_i, a, k_1, and k_2 are parameters, all positive except for k_2.

In order to understand the behavior of this oscillator, consider first the system of the above equations with the first equation truncated so that only the first three terms on the right hand side remain.

$$
\begin{aligned}
&\tau\,\mathrm{d}u_i/\mathrm{d}t = -u_i - \beta \cdot v_i + u_{\mathrm{e}}\,,\\
&\tau'\mathrm{d}v_i/\mathrm{d}t = -v_i - y_i\,.
\end{aligned}
\tag{8.35}
$$

The system obtained by fixing y_i in the second equation to either 0 or u_i is a system of second-order linear differential equations, and oscillates if the parameters τ, τ', and β are set appropriately. Since y_i switches between two values, the system oscillates, switching between two linear systems. The effect of the third term u_{e} in the right hand side of the first equation is to generate a limit-cycle, and, roughly speaking, an increase in this value results in an increase in the amplitude of the limit-cycle.

Returning to the original equations (8.34), the fourth term in the right hand side of the first equation connects the two subsystems. The term $\tilde{y}_i$ for mutual interactions is a variable whose value switches between 0 and u_i of the other subsystem, and it can be seen that, since setting y_i to u_i amounts to negative coupling of the subsystems, the two subsystems synchronize in antiphase.

The fifth term with s_i is for interactions with other oscillators. The effect of each subsystem is weighted with the same coefficient a, but their interaction is defined as follows:

$$
\begin{aligned}
&s_i(p) = \frac{2}{1+e^{-g_i(p)}} - 1,\\
&g_i(p) = \Sigma_q w_{pq} y_i(q),
\end{aligned}
\tag{8.36}
$$

where the oscillators' variables are given their own index p. If the value of y_i in the second equation is u_i, g_i is the linear interaction term as in the previous section. The first equation is the sigmoid function, which is one of the action models of neurons. Since the sigmoid function can be approximated by broken lines as shown in Fig. 8.17, the interaction by s_i causes synchronization just as in the previous case.

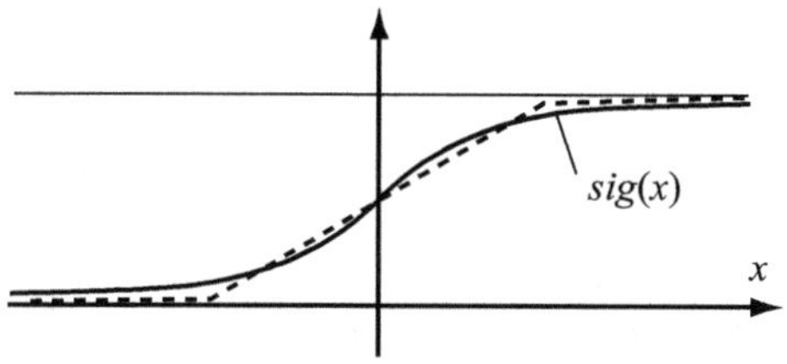

Fig. 8.17 Sigmoid function

The last term f_i is for the input from the physical system, e.g. a muscle, to the oscillator. In order to provide inputs to the two subsystems connected in antiphase, the effect of f_i in the first equation in (8.34) is multiplied by the coefficient k_i, which has the opposite sign.

When connecting oscillators to a physical system, if each joint actuator is composed of flexor and extensor as in biological organisms, y_1 and y_2 are connected to them separately. When controlling a conventional robot, each joint motor is driven by the output z, which is defined as $y_1 - y_2$.

Of course, in order to control robots, it is not necessary to simulate biological organisms. Other oscillators may work, such as a phase oscillator, a circular oscillator or a van der Pol oscillator. However, in the case of controlling M-TRAN robots, which we discuss later in this chapter, after trying various types of oscillators, Matsuoka's oscillator turned out to be the best. In the sections above, we focused only on synchronization, but for robotic motion control, not only the phase but also the amplitude function should be set appropriately[9].

8.4 Genetic Algorithm

Now that we have a general framework for generating rhythmic motions, in this section we consider how to use it for robot locomotion control. In Sections. 8.1.2 and 8.1.3, the coordinated motions were described by gaits or motion patterns, but

[9] Apart from M-TRAN, there are other modular robots that use CPG. The YaMoR robot mainly uses a circular oscillator [20], and the SlimeBot (Section 6.4) uses van der Pol oscillators [13].

in the case of CPG control, they are described in terms of oscillators' connections. Moreover, global entrainment has to be taken into account. As a result, designing a CPG locomotion control system becomes a problem of optimization, discovering oscillator connections suitable for locomotion by trial and error.

There are various optimization methods depending on the target. Generally speaking, asymptotic search methods that use derivatives, like the hill climbing method, are not effective where there are multiple local maxima or where the characteristics change drastically. A method called metaheuristics has been attracting attention as a solution to difficult search problems like these. The *genetic algorithm* (called GA hereafter) is one such method, which is now widely applied to various problems including complicated non-linear problems, combinatorial optimization, and NP-hard problems [20, 21].

GA is an optimization method that imitates the biological mechanism of evolution involving inheritance and selection (survival of the fittest). In GA, first we consider the system parameters as genes. When there are multiple parameters, a gene g can be regarded either as a vector composed of the parameters, or as a sequence of numbers or letters (Fig. 8.18(a)). A system that carries a particular gene, that is, a system whose parameters are specified, is called an *individual.*

Let us assume that there are some fixed numbers, n, of individuals with different genes. We call a collection of these individuals and equivalently a collection of genes, $G = (g_1, g_2, ..., g_n)$, a *generation*. From a generation, the next generation is obtained by a mechanism which simulates inheritance and selection (Fig. 8.18(b)). A single real number value, F, called *fitness* is assigned to each individual for the purpose of optimization. The fitness of every individual in the current generation is calculated, and the individuals of the next generation are determined by applying operations, i.e., *selection*, *mutation* and *crossover*, to individuals of the current generation according to their fitness.

While selection simulates the survival of the fittest by choosing individuals with higher fitness values, mutation is a process probabilistically changing part of a gene of a chosen individual. Crossover is a process of creating a new gene (and a child individual that carries it) that has traits of the genes of two chosen individuals (parent individuals). Many types of crossover have been proposed, but in the case of the M-TRAN robots the following two types are used:

- N point crossover: a gene is regarded as a sequence of letters. Each of two given genes is divided the same way into N segments. A sequence for a new gene is constructed by alternately joining segments from the two genes while preserving the original order. Fig. 8.18(c) is a description of two point crossover.
- Unimodal normal distribution crossover (UNDX): a method applicable to real number parameters. As described in Fig. 8.18(d), in the parameter space, the point of the child individual (g) is probabilistically determined using a normal distribution along the line connecting two parent individuals (the points g_i and g_j in the figure).

The procedure described above is repeated a specified number of times, and the individual with the highest fitness value is chosen from the last generation as the optimal solution.

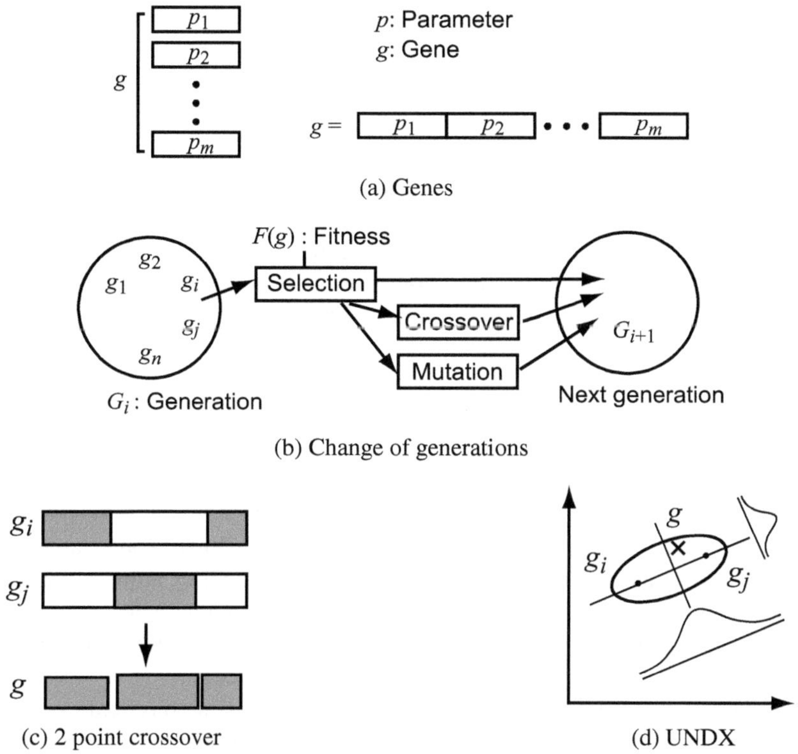

Fig. 8.18 Principles of genetic algorithms

8.5 Motion Control of the M-TRAN Robots[10]

The methods described in the previous sections, i.e., CPG motion control, the Matsuoka's oscillator, and GA, were applied to locomotion control of the M-TRAN robots.

8.5.1 CPG Control System

Given a robot configuration consisting of M-TRAN modules, consider a control system in which a Matsuoka's oscillator defined by (8.34) is connected to each of

[10] The content of this section is based on the research by Akiya Kamimura, AIST, Japan [22].

modules' joints (Fig. 8.19(a)). As there are cases, like the configuration in Fig. 8.19(b), in which the joints need to move around angles other than 0°, an offset angle θ_{i0} is set for each joint, and the difference between the actual joint angle and the offset, $(\theta_i - \theta_{i0})$, is used as an input to the oscillator. For n modules, $2n$ oscillators are used, and the interaction coefficient between each pair of oscillators w_{ij} is given. Similarly to the case discussed in Sec. 8.2, the oscillators' coupling is set to be symmetric, i.e., $w_{ij} = w_{ji}$, and thus the connections in both directions are represented by a single line in the following diagrams.

8.5.2 Fitness and Dynamics Simulation

Optimization by GA requires evaluation of the fitness function. Since our target is locomotion control, we define the fitness F so that an individual that travels further along a straight line while consuming less energy will have a higher fitness value:

$$F = aL - bD - cE\,, \tag{8.37}$$

where L and D represent, respectively, the distance in a straight line covered per unit time and the deviation from the straight line, E is the energy consumption per unit time, and a, b, c are positive weighting coefficients.

Evaluation of this fitness function F cannot be made analytically, and it is not practical to do it using a real robot, either. Since GA requires numerous trial-and-error attempts, it is a standard practice to use a numerical simulation for evaluation. A robotic motion, even static walking, is intrinsically dynamic, hence simulations need to be made by taking dynamics into account.

Dynamics simulation determines the behavior of a physical system by numerically calculating the integrals of equations of motion. Once the equations of motion are formulated, numerical integration can be easily done using algorithms such as Runge-Kutta method. But formulation of the equations of a mechanical system consisting of many parts connected in a complicated way is not a simple task. Issues such as geometric constraints of bearings and linkages, whether parts come in contact, and, in case of locomotion, friction and rebound due to contact with the ground need to be taken into account.

Recently, many generic packages have been developed for solving dynamics problems. By specifying physical and geometric parameters such as the mass, inertia moment, and three dimensional shape and connections of each component, they are capable of numerically computing the dynamic behavior of the system, even including the contacts and collisions, and displaying this three dimensionally. In our study, a commercial library (Vortex by Critical Math Labs) was used, but there are free packages available as well, such as ODE (Open Dynamics Engine) and OpenHRP (AIST) [23].

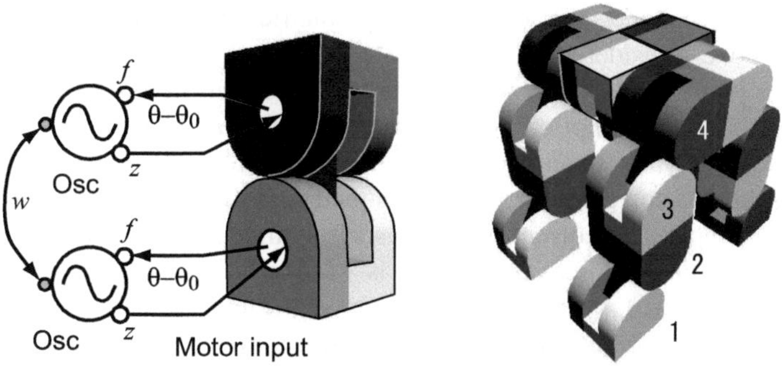

(a) CPG connection (b) Parallel quadruped form

Fig. 8.19 CPG control system and an example application

The dynamic characteristics of an M-TRAN module were modeled in two blocks, active and passive, with the same characteristics. Since servo control of joint angles is not appropriate for our CPG control, DC motor dynamics was also modeled. The friction with the ground was modeled as a combination of static friction and viscous friction. As a result, in the case of a robot consisting of n modules, in the system there are $4n$ variables for joint angles and their time derivatives, $8n$ variables describing oscillators for each joint, six variables describing the three dimensional position and the posture of the whole robot, and six variables as their time derivatives, so that the number of variables is $12n + 12$, apart from the DC motor variables.

Simulations were run under gravity starting from the initial state where all the joints are set their offset angles and the robot is resting on level ground. Fig. 8.20 shows examples of simulations using two joints and four joints. In the case of two joints as in Fig. 8.19(a), oscillations both in phase and in anti-phase (Fig. 8.20(a) and (b)) were obtained depending on the sign of the connection coefficients. In the case of linear configuration in Fig. 8.20(c), by connecting the three oscillators by negative weights, phase differences of roughly 120 degrees are obtained (Fig. 8.20(d)). These phase differences generate a traveling wave in modules arranged in a linear configuration, and it results in forward movement of this robot when it is placed on the ground.

8.5.3 GA Optimization

The design parameters for optimization, i.e., the genes, are the coupling weights w_{ij} of the oscillators. To reduce the GA search space, we only consider three values for each coefficient w_{ij}: 0, w, and $-w$ with a fixed value of w.

In the process of optimization, no prior assumptions are made. In the case of the quadruped configuration in Fig. 8.19(b), for example, it might seem appropriate to connect the oscillators symmetrically as shown in Fig. 8.21, because, as we considered in Section 8.1.2, it may be sufficient to set up an appropriate phase relation among the four legs and then to move each foot in the same way. If this assumption is correct, many of the connections can be set to have zero weight and several of the remaining connections can be set to have equal coefficients for the sake of the symmetry. However, there is no guarantee that this assumption is correct and that such symmetric connection leads to the optimal locomotion. Moreover, if there is even a little change in configuration, conditions based on symmetry cannot be fulfilled. We therefore made no assumptions for any configuration, started from a system which assumes that all joints are randomly connected, and then attempted to find connections appropriate for locomotion through an optimization process.

We considered additional design parameters, i.e., the initial state variables of the CPGs. A pair of harmonic oscillators following (8.18) that we discussed earlier have only one limit-cycle, but if several oscillators are connected in a loop, there can be two limit cycles, one in reverse order, as described in Fig. 8.13. Moreover, the above discussion of synchronization and entrainment dealt only with steady state behavior of the system. In the case of walking robots, even if there is an optimal CPG connection that realizes steady walking, the robot with such connection may stumble while in transition from the resting state to this steady state. In order to get a desirable transient behavior as well, we included the initial states of oscillators into the design parameters.

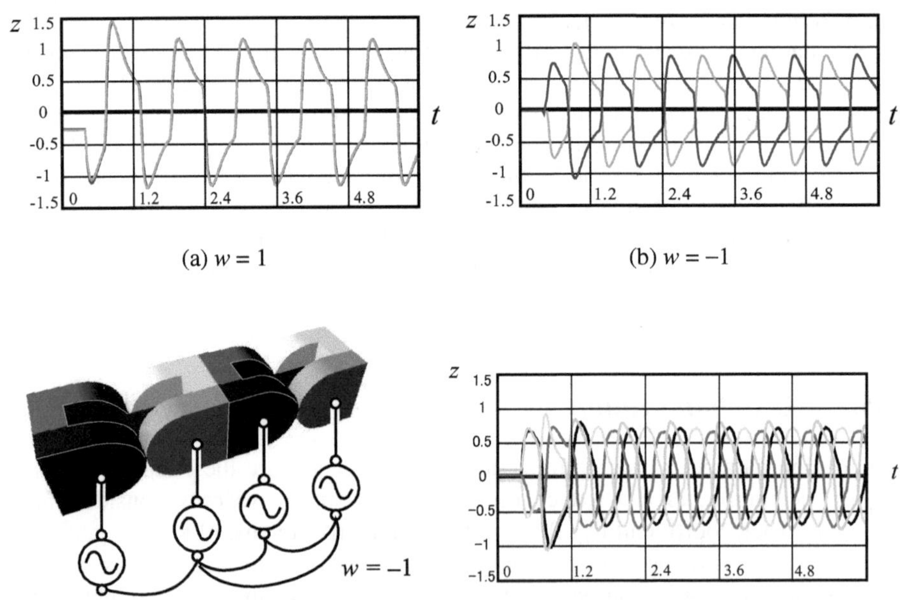

(a) $w = 1$ (b) $w = -1$

(c) Two module chain (d) Wave motion

Fig. 8.20 Example simulations. (a) and (b) are results for two joint case in Fig. 8.19(a).

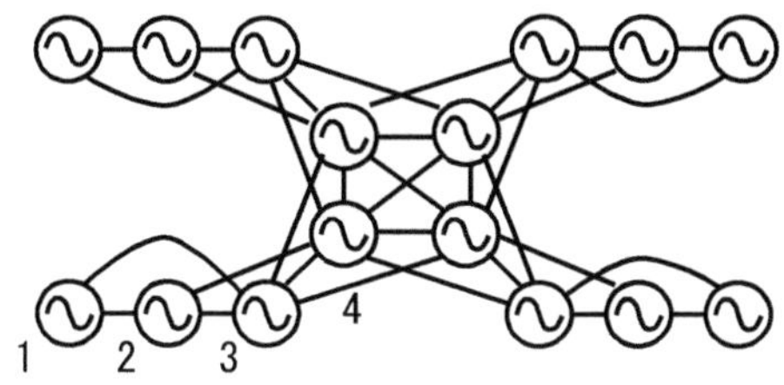

Fig. 8.21 Candidate connection for the configuration in Fig. 8.19(b).

Thus, for n modules, genes for GA are constructed from ${}_{2n}C_2$ variables with the discrete values 0, 1, and −1, and $8n$ real number values. Applicable methods for crossover and mutation when optimizing discrete values and when optimizing continuous values are different. We applied the N point crossover method for connection coefficients, which are discrete values, and UNDX for initial states, which are real number values.

8.5.4 Optimization Results and Playback Experiment

Fig. 8.22 shows one of the results of GA optimization. Various configurations as shown in Fig. 8.23 were also tried, and the optimization results were applied to the actual M-TRAN robots. The configurations in Fig. 8.22(a) and Fig. 8.23(c) are topologically the same, but for their locomotion, different offset angles (θ_{i0}) of joints were used. If the offset angles are set to those when standing on four legs, quadruped walking is achieved, while if they are all set to zero, the two rows of traveling waves are obtained.

To begin with, instead of mounting the CPG control system on the robot, we used playback control for hardware experiments. This means that each module was given a reference trajectory of joint angles resulting from a steady oscillation simulation as in Fig. 8.22(c), and the joint angles were maneuvered by servo controllers to follow this trajectory. Most of movements by robots in Figs. 8.22 and 8.23 are quasi-static, so playback control worked well.

8.5.5 Real Time CPG Control

Playback control does not work well if we change environmental conditions such as ground friction or the tilt of the ground. In order to take advantage of the local and global entrainment in this system to deal with such changes, it is necessary to drive each module by the same CPG controller as in the simulation. To achieve that, the CPU in each module carries out numerical integration of differential equations for each pair of oscillators, and state variables are exchanged between oscillators that are connected via the intra-module network bus. For this real time CPG control, optimized connections and oscillators' initial states are used to achieve the same motion as the simulations.

(a) Target robot configuration

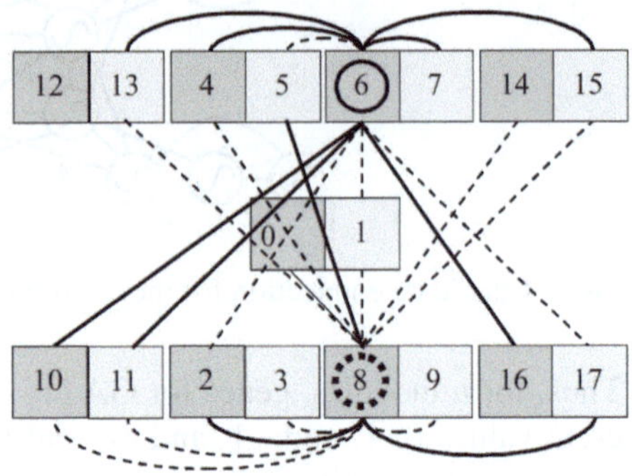

(b) Part of optimized connections

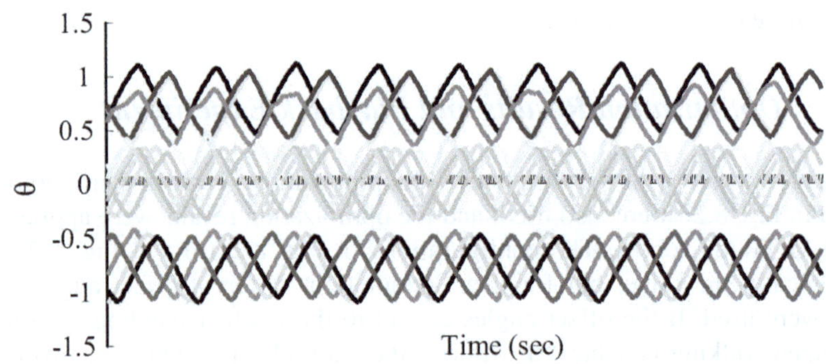

(c) Changes of joint angles

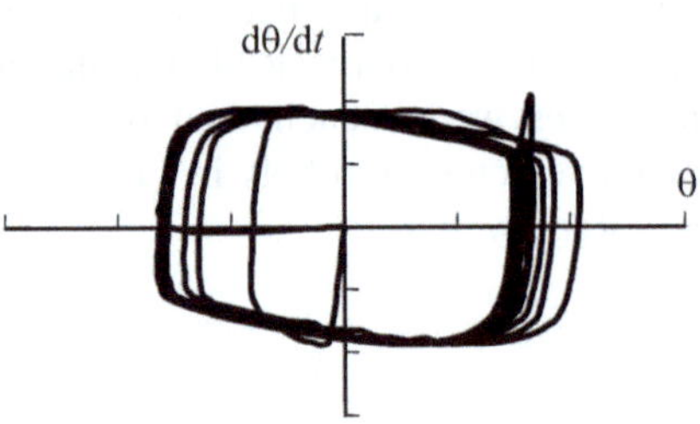

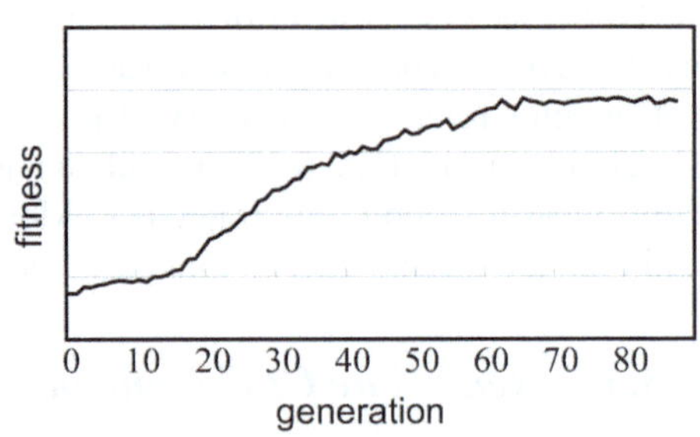

(d) Joint angle trajectory in the phase plane (e) Maximum fitness of succeeding generations

Fig. 8.22 GA optimization example. (a) A parallel quadruped robot, the design target. (b) Optimized coefficient of the oscillator coupling, for only the connections relating to the joints 6 and 8. Solid lines are for weight w and dash lines for $-w$. (c) Changes of joint angles. (d) A trajectory of a joint angle oscillation expressed in a phase plane. (e) The change in maximum fitness of each generation during GA computation

(a) Linear form with traveling wave

(b) Parallel six-legged form

(c) Two rows with traveling waves

(d) Minimum quadruped form

(e) Spider (four-legged)

Fig. 8.23 Robot configurations

In the rolling motion shown in Fig. 8.24, CPG control differs greatly from playback control. Since this motion is intrinsically dynamic and the playback control never synchronizes to the actual rolling motion, the robot was not able to maintain rolling motion and suffered excessive stress on its joints. By the real time CPG control, the rolling continued smoothly as seen in Fig. 8.24. It is notable that this CPG controller does not use a sensor specifically to measure rolling or a central program that determines whether the robot is properly rolling. The connection between each oscillator and each joint motor can be regarded as a proportional controller, in which the output from the oscillator causes motor torque and the joint angle value is input to the oscillator as feedback. Roughly speaking, this proportional control works as a sensor for external forces, and global entrainment works as feedback control of the whole system.

In the case of the quadruped robot in Fig. 8.22(a), a dynamic walk gait with two swing legs at a time is obtained. Even with this gait, the playback control worked well when the ground was flat and the actual friction coefficient was close to the value used in the simulation. However, when there was some change in the tilt of the ground or the friction coefficient, there were problems such as a swing leg

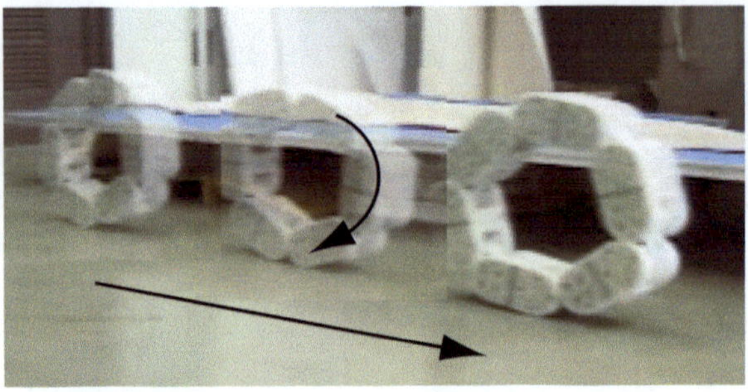

Fig. 8.24 Global entrainment during rolling motion

colliding with the ground, or slipping of a support leg. With the real time CPG control, on the other hand, the robot often was able to maintain its motion under such changes in the environment, by changing the stride or the pace.

Since the connection of each oscillator and each motor provides a proportional control, each joint has the characteristics of a spring. As with the serial linkage in Section 8.1.3.5 which uses ground friction and a traveling wave for locomotion, the serial M-TRAN robot under playback control could not go over the bump in Fig. 8.25. Under the real time CPG control, the robot successfully went over the bump using its spring characteristics.

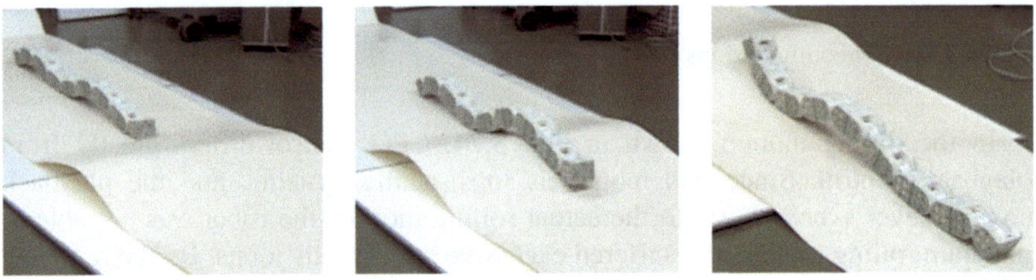

Fig. 8.25 Moving over bump using flexible traveling waves

Actually, the characteristics of proportional control may be either an advantage or a disadvantage depending on the situation. Integral compensation is a typical way to improve a proportional control system in many situations. The motion in Fig. 8.26 is an example of application of proportional integral control with its capability of disturbance estimation[11]. When the tilt of the ground is small (Fig. 8.26(a)), the proportional integral control is effective and four-legged walk is possible by adjusting the stride appropriately. When the tilt becomes steep

[11] In fact, the usage of the terms "proportional control" and "integral control" is not precise here because they are not directly applied to the joint angle control. In reality, they refer to proportional or integral control over the offsets of periodic movements.

(Fig. 8.26(c)), however, no adjustment makes it possible to climb by walking. This situation can be identified by estimating the disturbance with the integral feedback term. In the experiment, the robot changed its configuration into two rows when the estimated disturbance exceeds the threshold, and it continued climbing by traveling wave motion (Fig. 8.26(d)).

(a) Tilt angle 10 degrees

(b) Tilt angle 0 degree

(c) Tile angle 15 degrees

(d) Motion after metamorphosis

Fig. 8.26 Adaptive metamorphosis under disturbance

8.5.6 Issues of CPG Control

In the above system, the same modules and the same oscillators are used, and there is no central module that is in control of the whole system. It is indeed a homogeneous distributed system. Synchronization among modules is the result of the CPG control, and feedback of the entire motion also is utilized in the CPG control system. However, this is not the kind of a distributed system we saw in the previous chapters which uses local interactions only. As can be seen in Fig. 8.22(b), this system has connections even between oscillators of joints far apart from each other. Like the central nervous system of real biological organisms, such remote connections are indispensable, for dynamic walking in particular.

There are issues to be resolved in this CPG control of M-TRAN robots. We briefly discuss them below.

The CPG control in this chapter assumes movements only on a flat ground with uniform friction. In some experiments, the robot showed robustness against changes in friction and the tilt of the ground, but such robustness is not enough for autonomous motion over an actual terrain. Advanced recognition functions such as visual sensing must be incorporated. Integration of environmental recognition, decision making, and motion control, into a layered control hierarchy, whether using model-based control or CPG control, is still a difficult problem.

Various gaits are possible for multi-legged robots, as we explained in Section 8.1. Animals actually switch between different gaits depending on various factors[12]. In our study, however, only one gait has been obtained for each configuration so far. Future studies may include optimization with more parameters and with variation of oscillator functions.

8.6 Remark

The essence of distributed motion control lies in synchronization. In this chapter, we explained the mechanism of the CPG control that realizes synchronization and feedback control in the same framework, and, as an example of its application, various motions of M-TRAN robots were designed and experimented. It is difficult to convey by words and photographs that motions of the M-TRAN robots have smoothness and flexibility similar to biological organisms. We recommend that readers watch the videos available as extra materials.

The design process of motion control involves an initial setting of a homogeneous mechanism of control, and then repeated trial and error with simulations, finally achieving the optimal control system suitable for each robot. This is indeed a process of self-organization.

It is also feasible to make real robots learn motions by themselves using the real hardware instead of simulations, or using reinforcement learning instead of GA [19]. What is more, the behavior and the bodily form of all living creatures have been evolving in parallel. This is simultaneous optimization of structure and control, and adopting this scheme, a robot would evolve by changing both its physical configuration and controller structure [25, 26].

References

[1] Craig, J.J.: Introduction to robotics: mechanics and control, 2nd edn. Addison-Wesley (1989)

[2] Todd, D.J.: Walking machines: an introduction to legged robots. Kogan Page (1985)

[3] Sastra, J., Chitta, S., Yim, M.: Dynamic Rolling for a Modular Loop Robot. In: Adv. Robot., vol. 39, pp. 421–430. Springer (2008)

[4] Hirose, S., Morishima, A.: Design and Control of a Mobile Robot with an Articulated Body. Int. J. Robot. Res. 9(2), 99–114 (1990)

[5] Kimura, S., Tsuchiya, S., Suzuki, Y.: Decentralized Autonomous Mechanism for Fault-Tolerant Control of a Kinematically Redundant Manipulator. In: IEEE Int. Conf. Syst. Man. Cybern., vol. 3, pp. 2540–2545 (1995)

[6] Winfree, A.T.: Biological Rhythms and the Behavior of Populations of Coupled Oscillators. J. Theoret. Biol. 16, 15–42 (1967)

[12] Typical natural gaits of horses are, in the increasing order of speed, walk, trot and gallop. There is an experimental study showing that horses seem to use the most energy efficient gait [24].

[7] Strogatz, S.H., Stewart, I.: Coupled Oscillators and Biological Synchronazation. Sci. Am., 102–109 (December 1993)

[8] Pikovsky, A., Rosenblum, M., Kurths, J.: Synchronization: A Universal Concept in Nonlinear Sciences. Cambridge University (2001)

[9] Kokaji, S., Murata, S., Kurokawa, H., Tomita, K.: Clock synchronization mechanisms for a distributed autonomous system. J. Robot. Mechatron 8(5), 427–434 (1996)

[10] Kuramoto, Y.: Chemical Oscillations, Waves, and Turbulence. Springer, New York (1984)

[11] Guckenheimer, J., Holmes, P.: Nonlinear Oscillations, Dynamical Systems, and Bifurcations of Vector Fields. Springer (1983)

[12] Thompson, J.M.T., Stewart, H.B.: Nonlinear Dynamics and Chaos. John Wiley & Sons (1986)

[13] Shimizu, M., Ishiguro, A., Kawakatsu, T.: Slimebot: A Modular Robot That Exploits Emergent Phenomena. In: Proc. Int. Conf. Robot. Autom., pp. 2982–2987 (2005)

[14] Cohen, A.H., Holmes, P.J., Rand, R.H.: The nature of the coupling between segmental oscillators of the lamprey spinal generator for locomotion: a mathematical model. J. Math. Biol. 13, 345–369 (1982)

[15] Yuasa, H., Ito, M.: Coordination of many oscillators and generation of locomotory patterns. Biol. Cybern. 63, 177–184 (1990)

[16] Ijspeert, A.J.: Central pattern generators for locomotion control in animals and robots: a review. Neural Netw. 21(4), 642–653 (2008)

[17] Fukuoka, Y., Kimura, H., Cohen, A.H.: Adaptive Dynamic Walking of a Quadruped Robot on Irregular Terrain based on Biological Concepts. Int. J. Robot. Res. 22(3-4), 187–202 (2003)

[18] Matsuoka, K.: Mechanisms of frequency and pattern control in the neural rhythm generators. Biolog. Cybern. 56, 345–353 (1987)

[19] Marbach, D., Ijspeert, A.: Online Optimization of Modular Robot Locomotion. In: Proc. IEEE Int. Conf. Mechtron Automat., pp. 248–253 (2005)

[20] Holland, J.H.: Adaptation in Natural and Artificial Systems: An Introductory Analysis with Applications to Biology, Control, and Artificial Intelligence. MIT (1992)

[21] Goldberg, D.E.: Genetic Algorithms in Search, Optimization, and Machine Learning. Addison-Wesley (1989)

[22] Kamimura, A., Kurokawa, H., Yoshida, E., Murata, S., Tomita, K., Kokaji, S.: Automatic Locomotion Design and Experiments for a Modular Robotic System. IEEE/ASME Trans. Mechatron 10(3), 314–325 (2005)

[23] `http://www.openrtp.jp/openhrp3/en/index.html` (accessed July 05, 2011)

[24] Hoyt, D.F., Taylor, C.R.: Gait and the energetics of locomotion in horses. Nat. 292(16), 239–240 (1981)

[25] Sims, K.: Evolving Virtual Creatures. In: Proc. SIGGRAPH, pp. 15–22 (1994)

[26] Marbach, D., Ijspeert, A.J.: Co-Evolution of Configuration and Control for Homogenous Modular Robots. Intel. Auton. Syst. 8, 712–719 (2004)

Chapter 9
Hardware and Software

Abstract. In this chapter, we discuss the practical aspects of hardware and software designs of the M-TRAN system. In principle, behavior of lattice-type modular robots like M-TRAN can be understood to some extent without any physical experiments, instead investigating the algorithm itself, or making computer simulations. Also, motion control of a robot can be investigated without a real machine, if precise dynamics of the robot is modeled. However, methods and algorithms established on abstract models are meaningless unless they are physically realized as real robots and machines. When building a self-organizing mechanical system, we need to consider various issues of many different areas such as mechanical engineering, control engineering, robotics, and systems engineering. In this chapter, we introduce topics we studied while building the M-TRAN system.

9.1 Hardware

We have developed three generations of M-TRAN prototypes (Fig. 9.1). Table 9.1 and Fig. 9.2 summarize their specifications and system structures. The M-TRAN I module, which only has mechanisms for motion and connection, was developed into M-TRAN II, which has a control circuit and a battery together with sensors and wireless communication capability, and this was further developed into M-TRAN III, with capability for local communication [1-3].

9.1.1 Structure and Mechanism

An M-TRAN module consists of three parts: an active block, a passive block, and a link (Fig. 9.3). Fig. 9.4 shows the internal structure of M-TRAN II and III modules, and Fig. 9.5 shows all the structural and mechanical parts of an M-TRAN III module. To meet the requirements for weight reduction, accuracy of dimensions, and rigidity, the main structural parts were machined from bulk plastic materials. M-TRAN modules are remarkably smaller and lighter than other 3-D modules. For instance, a module of the Three Dimensional Universal Connection System, made from aluminum, is 275 mm in length and 6.8 kg in weight.

S. Murata and H. Kurokawa: Self-Organizing Robots, STAR 77, pp. 211–233.
springerlink.com

Fig. 9.1 Appearance of M-TRAN modules (from the left, I, II, and III)

Table 9.1 M-TRAN module specifications

	I	II		III
Length of block (mm)	66	60		65
Weight (g)	440	400		420
Number of CPUs	1	3		4
Inter-module communication method	-	LonWorks	CAN	CAN Serial adjacent communication
Control method	Control by host computer	Synchronous control & CPG motion control		Synchronous/asynchronous distributed control
Wireless communication	-	Command transmission from the host computer		Bidirectional wireless modem (Bluetooth)
Connection	Permanent magnet + SMA actuator			Mechanical latch
Sensors	-		Proximity sensors, acceleration sensors	
Battery	-	Lithium ion battery		Lithium polymer battery
Number of modules produced	10	20		50

Each module contains driving mechanisms for two joints and three active connecting faces, and so the choice of actuators and the designs of mechanisms were important. If we disregard its function as a chain-type modular robot, and limit the usage of the M-TRAN to that of a lattice-type modular robot capable of self-assembly and metamorphosis, discrete state control is sufficient for both the rotation of joints and connection. Namely, the joints only need be controlled to be in five different angles, 0°, ±45°, and ±90°, and the connection mechanisms need only two states, connected and disconnected.

However, it is not easy to find actuators which are capable of discrete motions with sufficient power. Hydraulic or pneumatic piston cylinders, solenoids, piezoelectric elements, and electrostatic actuators are not easy to use when power sources and circuits have to be in the module. Other typical devices fitted for small discrete actuators are electromagnets, permanent magnets, and shape memory alloy (SMA), but they have their problems as described in the next section. In the end, for M-TRAN III modules we chose geared DC motors for both joint drives and connection mechanisms, and used On/Off, bang-bang, and PI (proportional-integral) controllers.

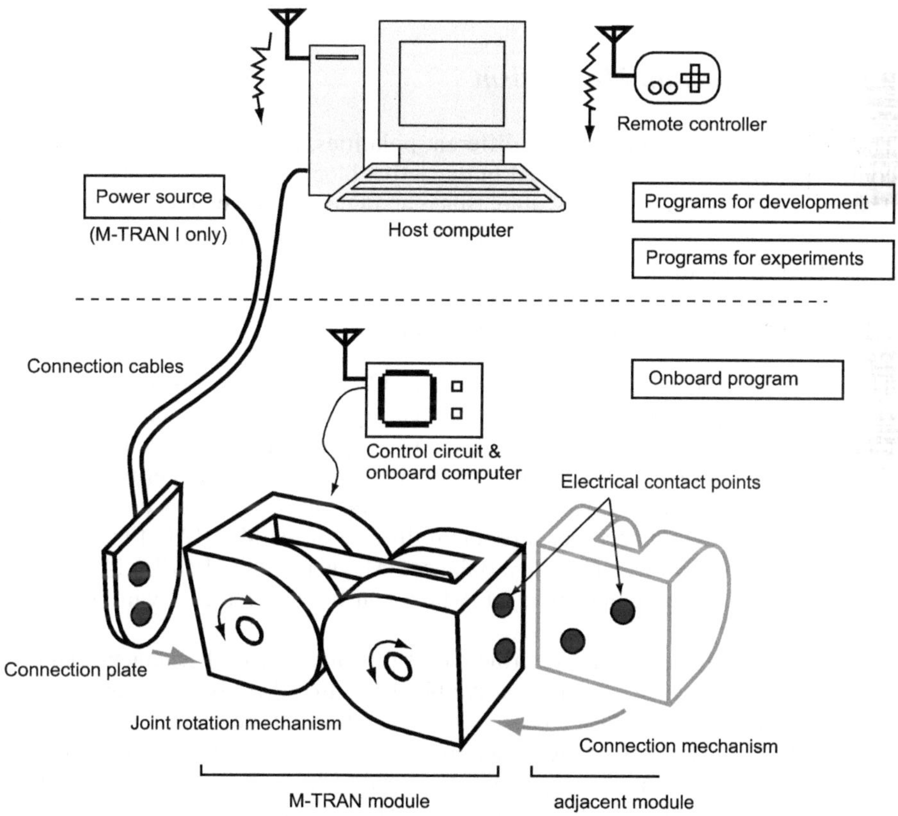

Fig. 9.2 Overview of the M-TRAN system

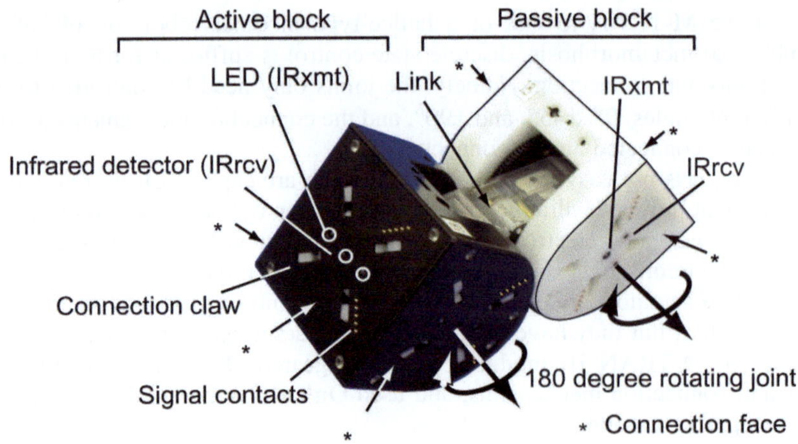

Fig. 9.3 External appearance of an M-TRAN III module

9.1.2 *Connection Mechanism*

The two blocks of a module have different polarities, and all the mechanisms for controlling connection are mounted in the active block, while the passive block is designed to be as simple as possible. Since a single module contains three active connection faces, a compact design of the connection mechanism is a significant factor in determining the size and weight of the entire module.

9.1.2.1 Magnetic Connection Mechanism

The connection mechanisms in M-TRAN I and II modules are magnetic. A simple connection mechanism can be made by pairing an electromagnet and an iron piece. This is controlled by switching the power on and off, but this is not a realistic method as it requires electricity to maintain the connection. A permanent magnet has the advantage that it maintains the connection without consuming power, but now disconnecting becomes a problem. In case of the M-TRAN I and II modules, a mechanism called Internally Balanced Magnetic Unit (IBMU) [4] is used, which makes a strong connection but which can be disconnected with a smaller force.

Fig. 9.6 explains the working principle of this mechanism. Permanent magnets are placed in both the active and passive blocks so that there is attraction between the blocks. The magnets of the passive block are fixed at the surface of the block, whereas those of the active block are fixed on a moving plate placed inside the block. This moving plate is supported by a spring which allows the plate to move up and down in the diagram. The plate receives several forces: the attraction F_m between magnets, repulsion F_s of the spring, the actuation force F_a, and the

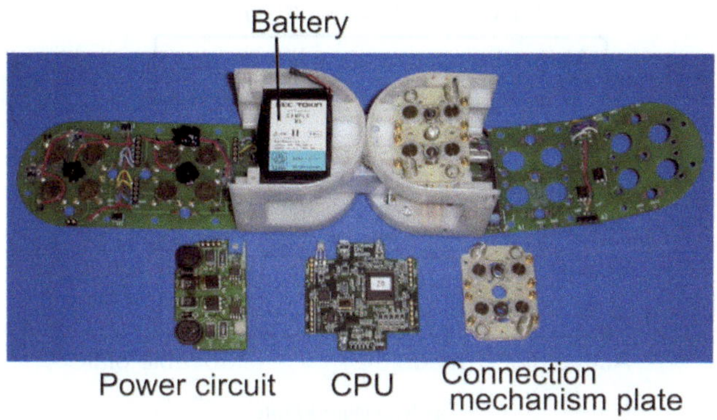

(a) M-TRAN II

(b) M-TRAN III

Fig. 9.4 Internal structure

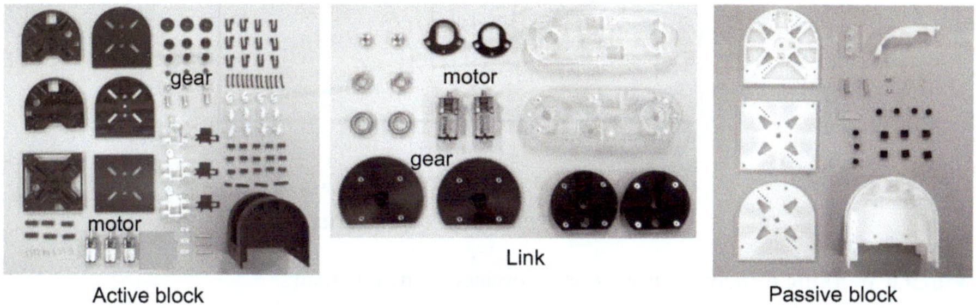

Fig. 9.5 Structural and mechanical parts of an M-TRAN module

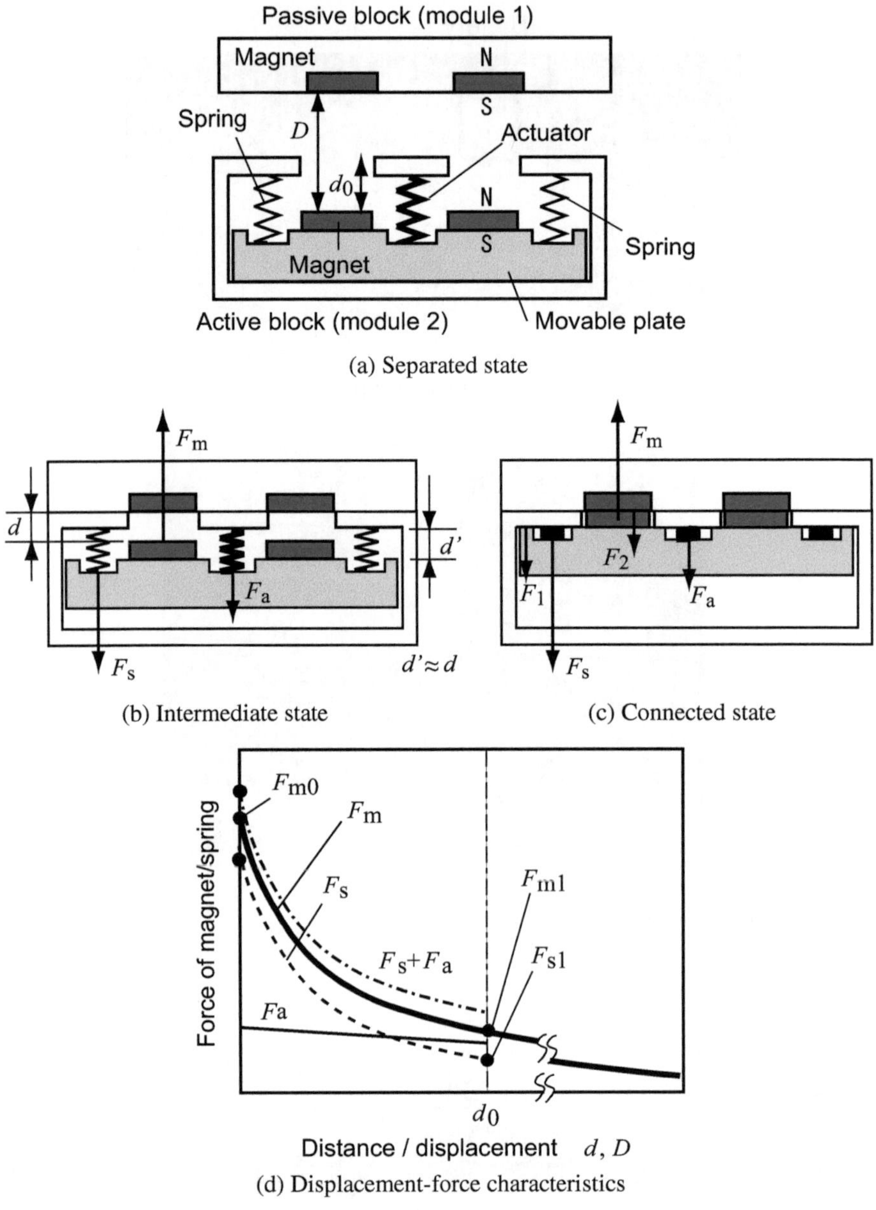

Fig. 9.6 Working principles of magnetic connection mechanism

counter forces F_1 and F_2 at points where there is contact in the connected state. Fig. 9.6(d) is a schematic graph that shows the strength of these forces with respect to the distance D or d between the magnets of the passive block and the moving plate.

The following two points should be noted here. First, since all the forces except for F_m and F_2 are inside the active block, regardless of the position of the moving plate, the connecting force (the force required to detach two modules) between the active and passive blocks is always F_m. Secondly, when two blocks are in contact as in Fig. 9.6(b), it is possible to make the plate move upward when the actuator's force F_a is 0 and move downward when F_a is slightly larger than F_m–F_s. This difference of forces, F_m–F_s, can be made sufficiently small for all d ($0 \leq d \leq d_0$) as in Fig. 9.6(d), by proper design of spring characteristics.

If the actuation force F_a is 0 and the two blocks contact each other, the movable plate in Fig. 9.6 (b) moves upward, and at d=0 the blocks are maintained in a connected state. In this situation, the connection force is the maximum value F_{m0}. To disconnect the blocks, the actuator is activated to push the plate downward (F_a in 9.6(d)). At the lowest point, the attraction force is F_{m1} at distance d_0, sufficiently smaller than F_{m0} for the modules to be disconnected easily.

As seen from the above, it is possible in principle to make the connecting force F_{m0} sufficiently large, and make the internal force F_a for releasing the connection and the external force F_{m1} for moving the modules apart sufficiently small. In reality though, the permanent magnets, non-linear springs, and actuators all require very careful production and adjustment. Also, the SMA (Shape memory alloy) actuator, used for M-TRAN I and II, was very slow and power consuming.

9.1.2.2 Mechanical Connection Mechanism

Mechanical connection mechanisms are used in several of the modular robot prototypes we introduced in Chapter 6, including the Molecule System, the Three Dimensional Universal Connection System, and the ATRON system. For the M-TRAN III, we installed a hooking mechanism (called connection hooks hereafter) in the active module, that clutches at the passive module [5]. M-TRAN modules differ from other prototypes in their capability to completely retract the connection hooks beneath the connection face. When hooks are retracted, two faces can slide along each other. Since the whole face of a block connects with another block, high rigidity is also achieved.

The connection mechanism consists of a motor, gears, linkages, sliding blocks, and connection hooks (Fig. 9.7). A connection hook is held by a moving axis on a sliding block and a fixed axis on the body (Fig. 9.8). The fixed axis moves in a groove in the connection hook, forming a cam mechanism so that when the sliding block moves, the hook first rotates about the moving axis and then slides, as shown in Fig. 9.8.

A connection face has four symmetrically-placed hooks which allow connections in the same way every time the face rotates 90°. The sliding blocks are driven by motors via linkages (Fig. 9.7). The link mechanism is designed so that when the faces are connected, the hooks are not retracted even when external forces are applied, allowing the connection to be retained without power. Also, the shapes of the hook and the cavities of a passive block are designed in such a way that the positioning errors are absorbed. The tolerance for positioning is 5 mm in a direction parallel to the face, 2 mm in the distance between faces, and 10 degrees of rotation of the face, for a face that is 65 mm square.

The connection is controlled using limit-switches. The time required for making and releasing connections is no more than five seconds.

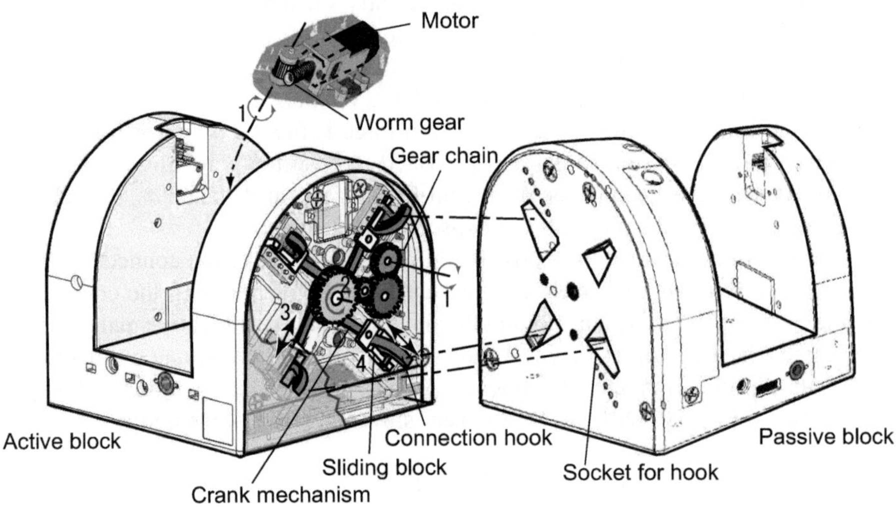

Fig. 9.7 Mechanical connection mechanism. This shows a mechanism of one of the three faces. Motion of the motor is transferred in the order indicated by the numbers to the sliding block, causing the straight movement of the moving axis.

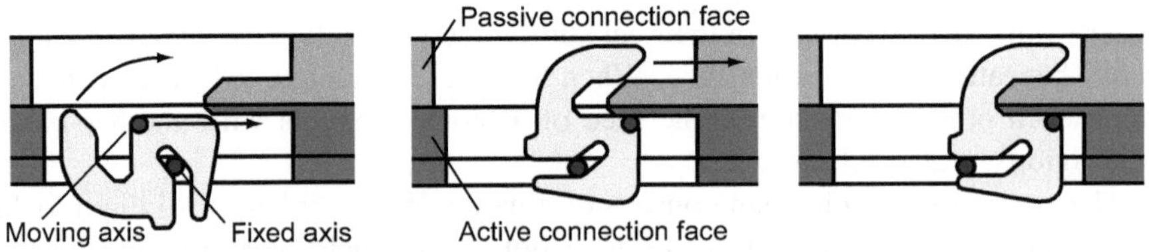

(a) Retracted connection claw (b) Straight movement starts (c) Completion of connection

Fig. 9.8 Action of connection hook

9.1.3 *Circuitry*

9.1.3.1 Multiple CPU System

Along with the advances in microcomputers and network technologies, recent control systems embedded in mechanical systems are generally designed in such a way that sensors and actuators are controlled by local CPUs, with these CPUs distributed in different locations being connected by a network bus. In the case of the M-TRAN system, the modules form a network, and also each module uses multiple CPUs, because multiple CPUs are easier to install than a single CPU with many wires for sensors and actuators.

The main CPU of an M-TRAN module controls the entire module, including three sub CPUs, one mounted on each of the three blocks (active, passive and link). Their connection and functions are illustrated in Fig. 9.9[1]. They share a serial signal line and are managed in master-slave manner (Fig. 9.9(1)).

A network bus called Controller Area Network (CAN) is used for communications among the main CPUs. The host computer is also connected to the CAN network as in Fig. 9.2, when the connection plate is attached to a face of a module. CAN is a bus network that uses two-wire balanced line. In order to use CAN between connected modules, each connection face has electrical contact points, so that when the modules are connected, the two bus lines become connected.

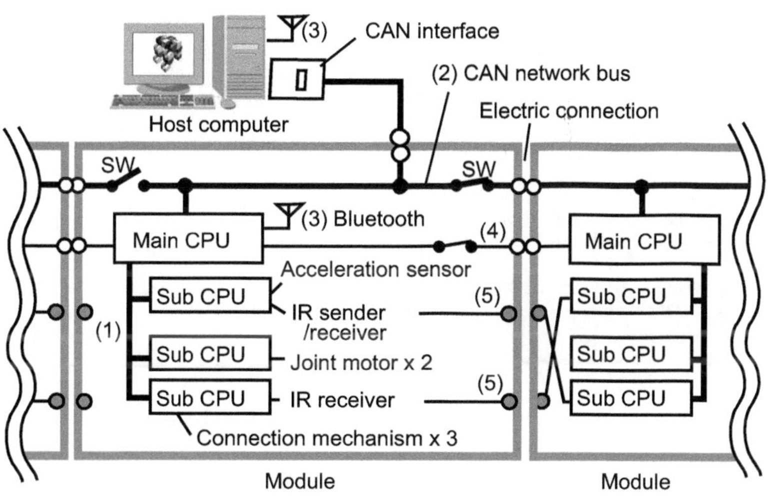

Fig. 9.9 Multiple CPU system. White circles between modules indicate electrical connections; black ones indicate non-contact connections.

[1] A Renesas 32 bit CPU SH-II (HD64F7047) is used as the main CPU. Two types of 16 bit CPUs (HD64F3687, HD64F3694) are used as sub CPUs.

Two sets of five electrical contacts are placed on every connection face of a module (Fig. 9.10). The positions of these contacts on the passive face are different from those on the active face so that in any orientation of the blocks, one of the sets are in contact. The five contacts in a set are: a common ground line, a pair of CAN bus lines (Fig. 9.9(2)), a connection detection signal line (Fig. 9.9(4)), and an additional common signal line. The common ground and signal lines are shared by all modules similarly to the CAN bus lines. The common signal line is used for starting and ending a program, and recovering from the sleep mode. One module has six connection detection signal lines for all six faces, and each of them is used for detecting the state of connection. All of these electrical contacts are placed on the module surface and therefore may cause short-circuit by contacting other modules or external objects. To avoid this, mechanical switches which are turned on only when there is connection with other modules are installed as in Fig. 9.9.

Apart from CAN communication, the host computer and the modules are capable of wireless communication over Bluetooth[2]. Also, there is serial infrared communication between adjacent modules (Fig. 9.9(5)). This is realized by a combination of an infra-red LED and a phototransistor, which are placed on connection faces in positions so that they face each other (Fig. 9.10).

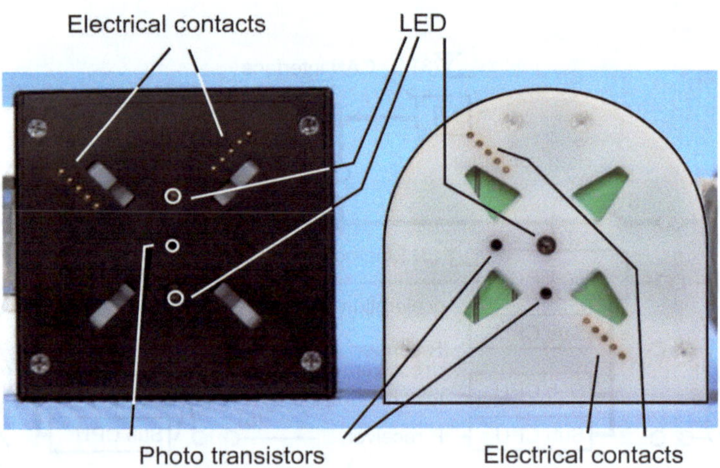

Fig. 9.10 Contacts and infrared devices

[2] M-TRAN III modules use Bluetooth for one-to-one modem communication, but this Bluetooth is not used to form a network.

9.1.3.2 Communication between Modules

Let us compare two types of communication: global bus communication such as CAN and Bluetooth, versus local adjacent module communication (Table 9.2).

Bus communications are high speed in general, but they require each module to have a unique identifier (ID). When the number of modules is large, the number of digits required for representing IDs increases. Moreover, since bus communications use shared bus lines, the more modules are used, the more frequently the messages collide and consequently the actual transmission speed decreases. On the other hand, communication between adjacent modules (by infrared etc.) is in general slower than bus communication, but since communication between each pair of connected faces is independent of other faces, the transmission speed is not affected by the number of modules used. Moreover, when infrared is used, there is no need for electrical contact.

Table 9.2 Comparison of communication methods

Global bus communication	Local adjacent module communication
Fast	Slow
ID required	ID not required
Relative coordinates of other modules unknown	Relative coordinates can be obtained by communication
Direct remote communication	Remote communication by relaying message
Genuine broadcasting	Broadcasting with time delay
Global Synchronization can be done by broadcast communication	Global synchronization is hard to attain
Speed is dependent on the number of modules, message collisions, traffic overload	Parallel communication, independent of module number

Using bus communication, it is possible for modules that are spatially distant from each other to communicate directly using their IDs, but a module cannot get information of its spatial position and orientation with respect to others through communication, and so it is indispensable to use an additional method for detecting spatial adjacency relations, such as the connection detection signal line. On the contrary, in the case of adjacent module communication, the message itself indicates the adjacency. When a connection face has two receivers and two emitters as shown in Fig. 9.10, relative orientation can be identified merely by message exchange. However, in order to exchange messages between remote modules using adjacent module communication, it is necessary to introduce an appropriate communication protocol that specifies the ways to describe message destinations, to choose a message path, and to avoid endless transmission of a message by broadcasting.

Thus, each type of communication has advantages and disadvantages. In motion control of chain-type modular robots, communication speed is the most important consideration and therefore bus communication is better. In fact, the PolyBot system (Fig. 6.23) uses the CAN bus, and the YaMoR system uses Bluetooth. On the other hand, for metamorphosis, scalability is more important than speed, which is one of the reasons why the ATRON system (Fig. 6.19), for example, uses infrared communication. In the case of the M-TRAN system, bus communication is used for centralized metamorphoses of configurations consisting of a relatively small, fixed number of modules, while adjacent module communication is used for distributed metamorphoses of regular structures of a variable number of modules. For CPG locomotion control, bus communication is used, and polling and broadcasting are carried out by a pre-selected leader to enable the modules to share state variables.

In M-TRAN III, it is possible to simultaneously transfer a program from the host PC to all modules via the connecting cables (Fig. 9.2). Generally speaking, in order to use onboard microcomputers, programs and data must be downloaded to each processor, and when large numbers of modules are used, this procedure is often very inefficient. Since all M-TRAN modules are designed to use the same

(a) With camera direction control

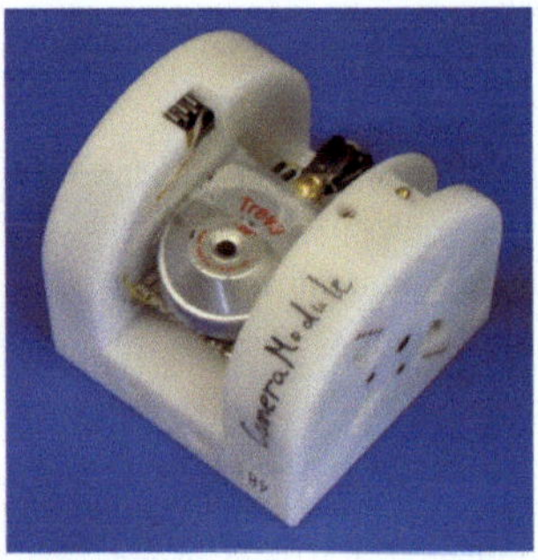

(b) With image processing capability

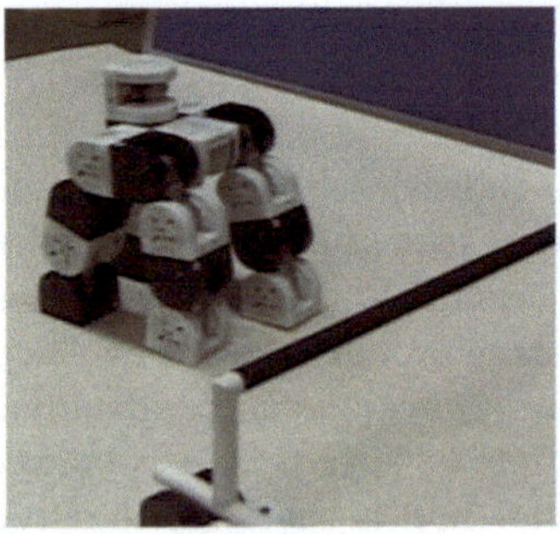

(c) Visual navigation

Fig. 9.11 Camera modules. In (c), the modules visually recognized the bar and changed their locomotion from walking to crawling program, downloading can be done all at once using this simultaneous transfer capability, which helps greatly in improving a development cycle of programming and experiments.

9.1.3.3 Power

Electricity is supplied to an M-TRAN I module via contact points on its connection faces. Connected modules share power when one module is connected externally to a power source via the connection plate as in Fig. 9.2. In such a configuration, the problem of voltage drop arises. For instance, when serially connected modules are provided with power at one end, the voltage drop on the other end is non-negligible, especially when the electric current is large, due to the accumulation of contact resistance at the intermediate contact points. Such a voltage drop is particularly problematic when connections or joint angles of many modules are controlled at the same time.

This problem is solved in M-TRAN II and III because each of these modules has a built-in battery. Since each module has a different level of battery consumption, it would be ideal if a module which has used up its own battery can receive power from other modules. Actually, such distributed power management is effective in large-scale systems such as an electric power transmission network. But in order to apply the idea to small-scale systems such as robots, we have to wait for small and highly efficient power control devices to become available.

9.1.4 Optional Modules

Modules with extra functions such as grasping hands or sensors can be made compatible to M-TRAN. Fig. 9.11 shows two types of camera modules. The module (a), which was used in the experiment in Section 7.4.2, is equipped with mechanisms for automatically stabilizing the camera direction and wireless image transmission to the host computer [6]. The module (b) is another camera module which uses the chassis, the power circuit, and the onboard CPU of M-TRAN III. Image processing is made by the onboard CPU, and the results are shared with other modules and/or transferred to the host computer via Bluetooth (Fig. 9.11(c)).

9.2 Software

Whatever the type of mechanical system, it requires a program for controlling the machine itself, and thus an environment for software development.

9.2.1 M-TRAN Simulator

One of the difficulties with designing three dimensional metamorphosis of M-TRAN is envisioning three dimensional transformations in your mind. Miniature blocks as in Fig. 9.12 are helpful, but the history of trials is hard to record, and it is not easy to track back your steps.

The M-TRAN simulator is a support tool that enables us to move and connect modules on a computer screen as if you are working with their models with your hands and eyes. It can store configuration data and the transformation processes (Fig. 9.13). The three dimensional display function was written in OpenGL, and GLUT (OpenGL toolkit) or Windows SDK were used for building the operation GUI.

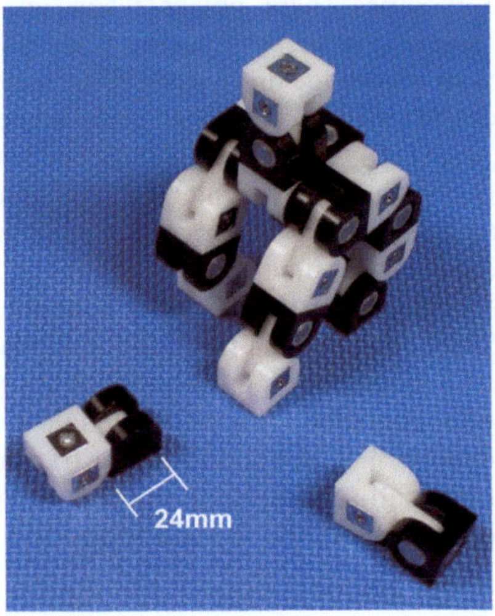

Fig. 9.12 Miniature blocks. Magnets are used for connection, and link angles can be changed freely.

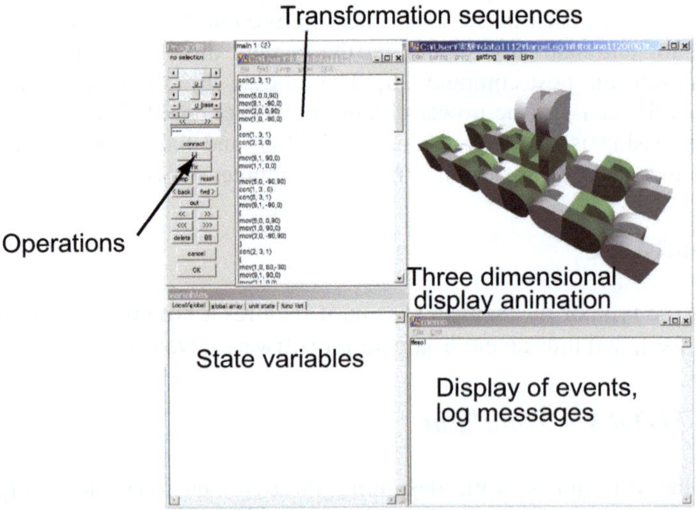

Fig. 9.13 Screenshot of development tool. Windows for displaying animations of kinematics simulations, the transformation program, state variables of each module, and operations.

The functions required of the M-TRAN simulator are computing and displaying results of three dimensional forward kinematics, and solving geometric problems such as collision and separation. In addition, the simulator can check stability of robots under gravity. Dynamic simulations provide detailed information on stability during motion.

A transformation procedure can be planned by choosing a module on the screen by a mouse and specifying changes in joint angles or module connections, selecting from among possible actions. Every operation applied to modules is displayed as a text, and they can be edited and saved. This text is a list of motion commands giving module IDs and parameters, and we call this a transformation sequence.

9.2.2 *Onboard Program*

Real time operating systems (real time OS) are effective in managing tasks with varying requirements and priorities, and are often used for computer control[3]. Nevertheless, we instead adopted a simple round-robin type method for the M-TRAN III, where all the tasks are called in order. This is because many tasks are either executed by sub-CPUs or have hardware interrupt functions.

The program running on the main CPU has a three layer structure. The bottom layer is in charge of CAN bus communication, timer interrupts, and program download via the CAN bus. The middle layer manages basic functions such as joint angle control, connection control and communications via CAN and infrared. The middle layer functions are included in the M-TRAN simulator so that the same source codes of the top layer can run either on the simulator or on the hardware. The top application layer includes main programs for transformation, either centralized or distributed, and uses the functions of the lower layer.

The middle layer is also equipped with a remote control mechanism over the CAN bus. The host computer can directly control modules via connection cables or Bluetooth, and also monitors CAN communication between modules.

9.2.3 *Program for Centralized Metamorphoses*

The various metamorphoses shown in Fig. 7.12 in Section 7.2.2 were realized by centralized control of a master-and-slave type. We here explain details of the control program, using the motion of a four-module minimum quadruped form in Fig. 9.14 as an example.

9.2.3.1 Transformation Procedure Data and Master-Slave Control

The transformation sequence for Fig. 9.14 is listed in Fig. 9.15. The same text data is stored by all the modules. Based on the data, master/slave type control is applied, where one module acts as a master and the remaining modules follow as slaves.

[3] SH-II (Renesas), the main CPU of an M-TRAN III module, is capable of running a real time OS called μITron.

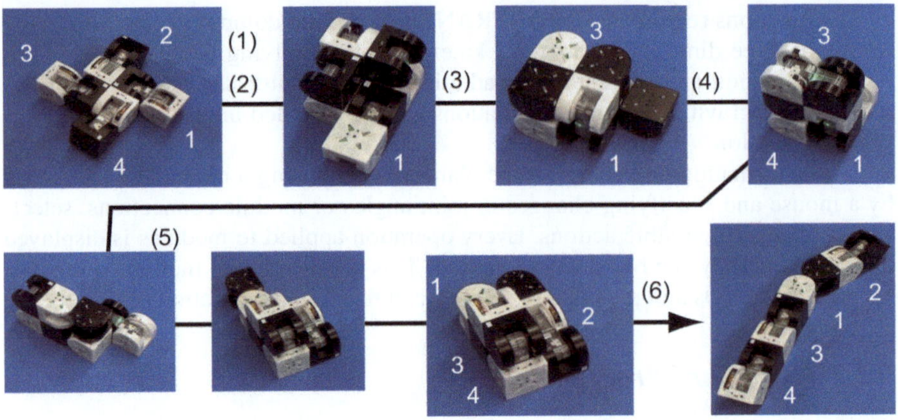

Fig. 9.14 Transformation procedure. Numbers in parentheses and those in white indicate the step number in Fig. 9.15 and the IDs, respectively.

Suppose that four modules with ID from 1 to 4 are assembled as in the first photo in Fig. 9.14, and that the module with ID 1 is set as the master. The master executes the data in Fig. 9.15 line by line. When the first two characters of a line are 'MV' or 'ML', the remaining characters are a triplet, (id, θ_a, θ_p), specifying joint motion. For a triplet with *id* 1, the master controls its own joint to (θ_a, θ_p), and for a triplet with another *id*, the master sends messages to the corresponding module for joint control. Each module sends a message back to the master when the joint control is completed. When the master receives all such completion messages, it moves to the next line of the transformation sequence. When the first two characters of the line are 'CN', the remaining characters are a triplet, (*id*, *fid*, *cntl*), in which *fid* (1 ... 6) specifies a connection face and *cntl* (0 or 1) specifies an action, either connecting or releasing. Connection of multiple modules is managed in the same way as the above joint control [3].

Experiments were carried out for various transformations shown in Fig. 7.12, by designing transformation sequences similar to that in Fig. 9.15. For each configuration and corresponding transformation sequence, a centralized control with parallel movements of multiple modules was realized as described above.

9.2.3.2 Configuration Recognition and Role Assignment

When arbitrary modules are assembled into one of the predefined configurations, modules need to recognize their configuration and the way they are assembled. Each module carries the same program and all the transformation sequences, so any module can be a master module.

The simplest algorithm for selection of a master (leader election) is to choose the module with the smallest ID. Another method is to select a module with a specific connection topology. During the leader election process, the number of

```
//---------Step 1---------------------------
ML, 1,-90,0, 2,0,90, 3,-90,0, 4,0,90,
MV, 2,-90,90, 4,-90,90, 1,-90,90,
CN, 2,2,1, 4,1,1,
//---------Step 2---------------------------
CN, 1,6,0, 3,2,0,
ML, 2,-90,45, 1,-60,75,
ML, 2,-45,45,
MV, 2,0,90,
CN, 3,2,1,
//---------Step 3---------------------------
CN, 2,2,0,
ML, 2,90,90, 1,0,90,
ML, 4,-90,0,
MV, 1,-90,90, 3,-90,90,
CN, 1,2,1,
//---------Step 4---------------------------
CN, 4,1,0,
ML, 4,0,90,
ML, 3,0,90,
MV, 2,-90,90, 4,-90,0,
CN, 2,1,1, 4,2,1,
//---------Step 5---------------------------
CN, 3,2,0, 3,4,0,
ML, 4,-90,-90,
ML, 4,0,0,
ML, 1,80,90, 2,-90,45, 4,-45,-90,
MV, 1,90,90, 2,0,-90, 4,-90,-90,
MV, 3,-90,90,
CN, 3,2,1,
//---------Step 6---------------------------
CN, 2,1,0
MV, 1,0,0, 2,0,0, 3,0,0, 4,0,0
```

Fig. 9.15 Transformation procedure. The actual data used in the experiment shown in Fig. 9.14. ML and MV at the beginning of a line indicate that the line specifies a change in joint angles (ML specifies rough positioning). The following three numbers specify the module ID (from 1 to 4) and two joint angles. CN indicates a change in connections, and the three numbers that follow specify the module ID, the connection face (1-6) expressed as a six bit number (000001 - 100000), and whether the face is to be connected (1) or released (0). The actions given in one line are executed by multiple modules in parallel, and when all the modules complete their execution, the next line is executed. The transformation is divided into 6 steps, and each step consists of disconnection (except for the step 1), changing joint angles, and making connections in that order. Some of the angle changes are not in units of 45°, for practical reasons.

modules and their IDs are obtained[4]. Next, the process of module configuration recognition is carried out by correlating the adjacency relations of all the modules with those of candidate configurations[5]. Once the configuration is recognized, the role of each module is assigned. Each module's ID is set to that for its place in the designed configuration (we call this the design model hereafter). Then, each module's orientation is checked as follows.

9.2.3.3 Symmetric Conversion

The adjacency relation between two modules is expressed by a pair of numbers indicating the faces in contact and orientation with respect to the local coordinates defined in Fig. 9.16. When modules are assembled with some freedom, some modules may have a different orientation than that in the design model.

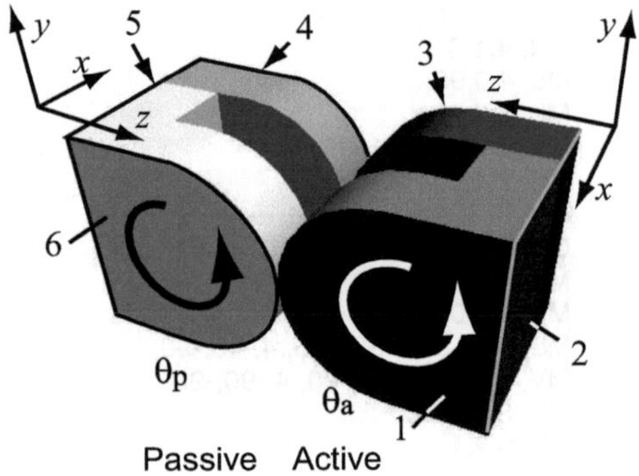

Fig. 9.16 Module coordinates and variables

This freedom in configuration is due to the symmetric transformations of a single module or of all the modules. There are three possible types of symmetric transformation: 180° rotation of each module around the z axis, the mirror image of the whole configuration, and switching between the passive blocks and the active blocks of all modules in the configuration (polarity conversion) (Fig. 9.17)[6]. These three transformations are equivalent to the coordinate transformations of

[4] In the explanation of distributed algorithms in Chapter 5, we mentioned that it is hard to know the number of all the processes, but in the case here, it is easy to detect the presence of a module and to know the total number because the range of module IDs are limited from 1 to 50, and broadcast using the CAN bus is available.

[5] This matching process can be done by individual modules in a distributed parallel manner, but this is equivalent to the centralized process.

[6] Vertical flip of the whole configuration is another geometrical transformation that retains local connective relations. As transformation experiments are made on the ground, though, vertical flip often makes a transformation unsuccessful.

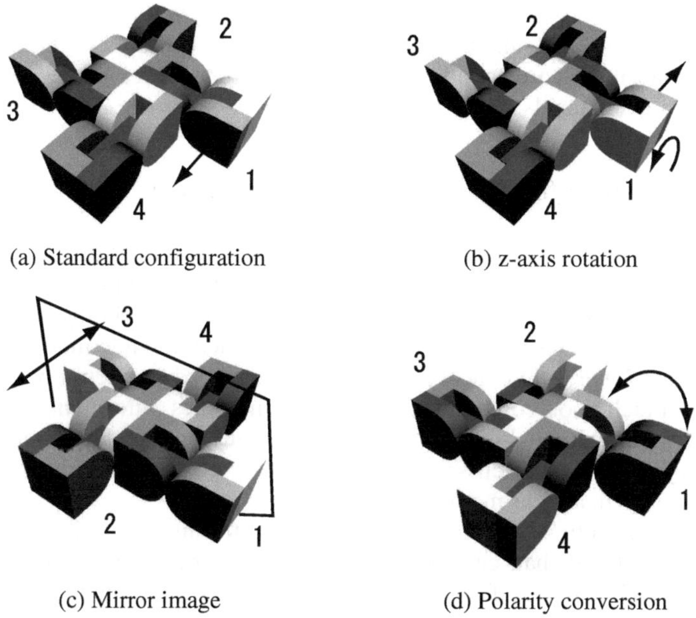

Fig. 9.17 Symmetry conversions

each module and cause changes in joint angles and face IDs as described in the list below. Therefore, when the configuration of the design model is transformed by either of these transformations, the corresponding change of joint angles and face IDs applied to a metamorphosis sequence results in the same metamorphosis:

[rotation around the z axis]
$(\theta_a, \theta_p) \rightarrow (-\theta_a, -\theta_p)$, *fid* :(1,2,3,4,5,6) → (3,2,1,6,5,4)
[mirror image]
$(\theta_a, \theta_p) \rightarrow (\theta_a, \theta_p)$, *fid* :(1,2,3,4,5,6) → (3,2,1,6,5,4)
[polarity conversion]
$(\theta_a, \theta_p) \rightarrow (\theta_p, \theta_a)$, *fid* :(1,2,3,4,5,6) → (4,5,6,1,2,3)

A three bit parameter corresponding to these conversions is set for each module during the process of configuration recognition. By adjusting the functions for joint angle control and connection control corresponding to the conversions, all the actions of a transformation sequence become exactly the same regardless of orientation of each module.

The polarity conversion, however, requires an additional mechanism. Since the connection is controlled only by active blocks, the command triplet (*id*, *fid*, *cntl*) in the transformation sequence is effective only where *fid* is 1, 2, or 3. When it is 4, 5, or 6, the module cannot initiate connection. In such a case, the module can remotely control the corresponding active face by sending a message via infra red communication channels.

9.2.4 Program for Distributed Metamorphosis

Distributed metamorphoses of regular structures discussed in Section 7.3 use only local adjacent module communication. The top layer of the three layered program is further divided into two layers, a layer for executing algorithms for an abstract model of meta-modules and the other for realizing meta-module motions.

In Chapter 7, we were concerned with kinematics and geometry, but we did not touch very much upon communication between modules, information sharing and synchronization. Although there is a leader in a meta-module, the problems of distributed algorithms we discussed in Chapter 4 have to be taken into account even among modules in a meta-module.

Generally speaking, program development and testing of distributed systems is much more difficult than centralized sequential programs. In debugging, for instance, it would be useful if distributed programs can be executed in a simulator in the same way as in a real machine. This entails parallel processing to simulate distributed behavior on a single computer. One efficient way to do so is to make a program for a single module, and to run its multiple copies in parallel as program threads[7].

Top layer programs with the same source code run on the simulator as threads and in the modules. Running a program in a different computer environment from the one it is written for is called emulation, and a circuit or an environment for this purpose is called an emulator. A perfect emulation is of course desirable, but development of an emulator itself is a challenge that requires time and effort. In developing a program capable of running on real machines, balanced development of widely applicable distributed algorithms, module programs, and emulators using abstract models is important.

9.3 Errors and Reliability

In Chapter 1, we stated that self-organization can improve the reliability of machines. Having said that, manufacturing errors and control errors are inevitable for not only the M-TRAN system but machines in general, and the larger the number of modules used, the more problems with reliability may arise. In this section, we outline typical errors and malfunctions of modular robots.

---

[7] We used Borland C++ on Windows for developing the simulator. The program for controlling modules was made using the TThread class, and as many instances were generated as the number of the modules, which run in parallel.

9.3.1 Dimension Error

In production of hardware, manufacturing errors in dimensions and angles always occur, and these may be compounded by control errors when the system is operated. In general production of machines, in addition to improving the precision of individual parts, parts are selected so that the error in one compensates for the error in another. In the case of self-organizing mechanical systems, however, all the modules are expected to be exactly the same, which may cause problems. For instance, placing another module in the gap in the middle of the configuration shown in Fig. 9.18 is possible in the design, but may not be possible in reality due to dimension error.

The problem of such dimension errors is common to most modular robots. To alleviate the influence of errors when building structures, it is necessary to give certain flexibility and freedom to the components. The Fractum modules discussed in Chapter 6 have flexibility in their magnetic connections, and the planar regular structure in Section 7.3 has internal gaps, resulting in some structural freedom.

9.3.2 Structural Deformation

Flexibility of components and connection, on the other hand, causes the problem of structural deformation due to external forces such as gravity. Fig. 9.19 shows an example of configurations that are prone to connection failures. Correction of such an error usually requires measurement and control in six degrees of freedom. However, the M-TRAN system does not have sensors for measuring such errors. Even if an error is measured, there may be less number of joints than necessary in the configuration, or some of the joints may be at their limits of ±90° and cannot rotate further. Therefore, alignment correction may often be impossible even by multi module cooperation.

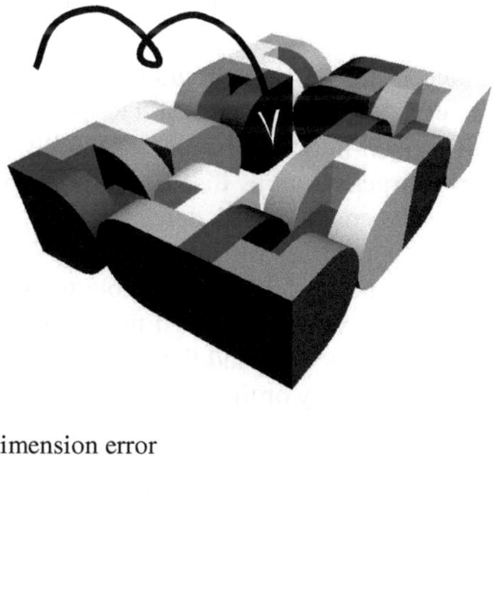

Fig. 9.18 Problem of dimension error

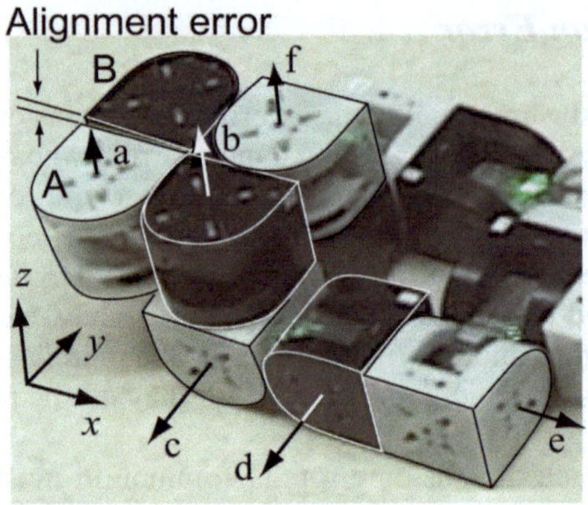

Fig. 9.19 Structural deformation and positioning. Though the blocks A and B are misaligned, rotation about the axes a and b cannot resolve the error by movement in the positive direction of the y axis, and rotation about the axis c is not possible in one direction, because they are at their angle limits.

9.3.3 Dealing with Errors

Instead of feedback control requiring coordination of modules, a simpler method at the single component level is preferred. Failures and errors as above should be compensated for in such a way that having enormous numbers of components is not a disadvantage but an advantage.

In the example shown in Fig. 9.18, for instance, which fails due to dimensional error, one solution is to give up on the current module and try again using another module. Also, in the experiments, we avoided the situation described in Fig. 9.19 by using another transformation procedure which caused less structural deformation. As there are several transformation procedures that yield the same or similar consequence, it is possible to choose a better procedure by trial-and-error, even with a real machine.

Furthermore, the lattice structure itself should be reconsidered. The lattice-type modular robots in Chapter 5, from Fractum to M-TRAN, assume idealized crystalline structures. Though a small number of modules readily form ideal structures, when the number is very large, errors as discussed above are inevitable. In contrast, natural crystals usually contain local defects varying from its ideal lattice structure. Indeed, the circular modules of SlimeBot (Fig. 6.27) can make connections at any point along their perimeters, and are able to create a planar structure without being constrained by any lattice, though they cannot make a lattice structure easily [7]. For large scale structures and their transformation, it may be necessary to change adaptively the rigidity or the flexibility of modules' connection.

References

[1] Murata, S., Yoshida, E., Kamimura, A., Kurokawa, H., Tomita, K., Kokaji, S.: M-TRAN: Self-Reconfigurable Modular Robotic System. IEEE/ASME Trans. Mechatron 7(4), 431–441 (2002)
[2] Kamimura, A., Kurokawa, H., Yoshida, E., Murata, S., Tomita, K., Kokaji, S.: Automatic Locomotion Design and Experiments for a Modular Robotic System. IEEE/ASME Trans. Mechatron 10(3), 314–325 (2005)
[3] Kurokawa, H., Tomita, K., Kamimura, A., Kokaji, S., Hasuo, T., Murata, S.: Distributed Self-reconfiguration of M-TRAN III Modular Robotic System. Intl. J. Robot. Res. 27(3-4), 373–386 (2008)
[4] Hirose, S., Imazato, M., Kudo, Y., Umetani, Y.: Internally-balanced magnet unit. Adv. Robot. 1(3), 225–242 (1986)
[5] Terada, Y., Murata, S.: Automatic Assembly System for a Large-Scale Modular Structure. In: Proc. IEEE/RSJ Int. Conf. Intel. Robot. Syst. (IROS), pp. 2349–2355 (2004)
[6] Murata, S., Kakomura, K., Kurokawa, H.: Toward a scalable modular robotic system – Navigation, docking, and integration of M-TRAN. IEEE Robot. Automat Magazine 14(4), 56–63 (2007)
[7] Ishiguro, A., Shimizu, M., Kawakatsu, T.: A modular robot that exhibits amoebic locomotion. Robot. Auton. Syst. 54(8), 641–650 (2006)

Chapter 10
The Future of Self-Organizing Robots

Abstract. In the chapters so far, we have discussed in detail the design of mechanical systems and robots based on self-organization, from their historical development to specific design examples. The authors hope that these chapters enable the reader to understand that these are fundamentally multiple-module systems that have flexibility that cannot be expected of traditional mechanical systems. In this chapter, we discuss the limitations and future challenges facing self-organizing robots built with electro-mechanical technologies, and then, returning to the philosophy of "design based on self-organization" one last time, consider the goals we should set for self-organizing robots from a relatively long term perspective. One such goal is to make *molecular* machines that self-organize.

10.1 Challenges for Self-Organizing Robots

As we saw in Chapter 6, there are many prototypes of modular robots other than M-TRAN. Although each of them has specially designed modules and their own target to achieve, they share the same aspect that they are built based on mechatronics[1], which means that they are built by combining existing devices so that the latest technologies can be readily incorporated. Viewed from another perspective, however, this also means that the performance of these systems is limited by the currently available devices.

10.1.1 Module Size

The speed and the density of microprocessors have been steadily improving as described by Moore's Law[2]. This though is an exception, and there are bottlenecks in the improvement of many other devices, such as motors and sensors, in both speed and size. The most notable limitation in a self-organizing mechanical system resulting from the use of such devices is the size of a module. With smaller

[1] Mechatronics is an interdisciplinary research area which is a combination of mechanical engineering and electronics. The term mechatronics was originally coined in Japan.

[2] An empirical rule that the density of transistors on an integrated circuit doubles every 18 to 24 months. It is said that we are approaching the end of this trend.

S. Murata and H. Kurokawa: Self-Organizing Robots, STAR 77, pp. 235–246.
springerlink.com

modules, the shape and the functions of a system can be specified with finer resolution, and also the power-to-weight ratio naturally tends to improve.

However, there is a limit to how small a module can be made. In the case of the M-TRAN for example, the module size is constrained by the axis length of the motor shaft. It is likely that the current length 60mm can be halved, but when this is achieved, the size of the battery and the electric circuit board will become critical in turn. In order to achieve module size of less than 10mm, we need to wait for downsizing of all component devices[3].

10.1.2 Number of Modules

The number of modules that can be used in one self-organizing mechanical system is limited by various constraints. For the systems developed so far, at most 100 modules have been manufactured (50 in the case of the M-TRAN modules and 100 for the ATRON modules).

This is partly because of the high production cost of a module, but essentially because of limited module reliability, in particular, reliability of connection mechanisms and of the communication channels and power supply lines that are established at the same time as the mechanical connections. When connections are made via electrical contacts, noise and voltage drops have to be dealt with, and as the number of modules increases, the system must be designed under the assumption that errors occur in both the physical and the communication connections.

Moreover, if the algorithm is to be given sophisticated functions such as automatic recovery from an error, the difficulty of software development, programming and debugging increases as well. One possible solution to such problems is to extend algorithms like the cellular automata we discussed in Section 7.3 to be able to deal with such probabilistic errors.

10.1.3 Choice between Self-reconfiguration and Self-assembly

M-TRAN and almost all the other self-organizing mechanical systems and robots that we discussed in this book are of the type where the modules are already connected in the initial state, and the reconfigurations of the whole system are realized by incremental changes in the connections. This method is advantageous in that the degrees of freedom required of a connection mechanism can be reduced, although it is very difficult to build a system out of randomly scattered modules in this way.

There is a very different approach to self-organization, in which a system is assembled through random collisions among parts, induced by agitating the container in which all the parts are placed, as we discussed at the end of Section 3.4.

---

[3] The photolithography technology allows building of circuits that are only a few micrometers in size on a silicon substrate. However, it is unlikely at the moment that we can devise a way to implement connection changing mechanisms and independent power sources of that scale in modules.

A two dimensional system like this may be plausible if methods like air flotation are used, but in three dimensions, modules of normal sizes are unlikely to form into the desired system[4]. However, since the supply of modules is a fundamental problem in self-repair or self-assembly, having modules floating in space and always available at hand is much to be desired.

10.2 From Mechatronics to Molecular Machines

Roughly speaking, the size of biological cells is about 2 μm in the case of bacteria and 20 μm in multicellular organisms. The components of the system of a cell are molecules, and these components can be obtained simply by catching them from the ambient solution, since molecules move randomly in a solution due to thermal agitation. Indeed, cells are even capable of reproducing themselves by taking in molecules that provide energy or serve as their structural parts. A multicellular organism is an assembly of an innumerable number of such cells. It is said that a human consists of around sixty trillion cells, a cluster of an astronomical number. It is truly amazing that an assembly of this scale can function in perfect coordination.

There is a fundamental difference between the ways of building electromechanical systems and biological systems. Machines and electrical circuits are made through the process of cutting away unnecessary parts from a lump of the metal to be used (processing by removal). So, unlike a single atom that functions by itself, a bulk material only has function after being processed. On the other hand, in the case of biological organisms, each single molecule has its own function. The molecules which are necessary are synthesized by chemical reactions when necessary and in the necessary amount, in a very efficient way.

Typical protein molecules are several nanometers in diameter, and in the case of E. coli, which is one of the simplest bacteria, more than four thousand kinds of such proteins are contained in a cell of about 2 μm in size. Some of those proteins, such as enzymes, function independently by themselves, while others function when many of these molecules assemble themselves into a certain structure such as a microtubule. Moreover, compared with semiconductors in which elements are aligned only two dimensionally, biological systems have very high space efficiency because the three dimensional space inside a cell is filled with various molecules.

It had been considered that molecular machines composed of biopolymers such as proteins and nucleic acids are unique to biological organisms, but recently, systems that are made from artificially synthesized molecules and that have functions similar to those of biological cells have been successfully created. If artificially synthesized cells with functions similar to those of biological cells are realized, self-organizing mechanical systems will have come closer to achieving both their ultimate structure and function, considering that they were originally inspired by biological organisms.

Below we discuss attempts to construct systems using such artificial molecules.

---

[4] In a stirring attempt conducted at Cornell, though, a large number of mechatronic modules were placed in a silicon oil vessel and stirred [1].

10.2.1 Molecular Machines Based on DNA Nanotechnology

The concept of nanometer scale mechanical systems first appeared in the famous lecture "There is plenty of room at the bottom" in 1959 by Richard Feynman at the American Physical Society [2]. In this lecture, he argued for the importance of research on devising ways for manipulating substances at the atomic and molecular levels, pointing out that it is possible to write all the information contained in the Encyclopaedia Britannica on the point of a pin without breaking laws of physics if it is possible to make each individual atom carry data.

Inspired by this idea, K. Eric Drexler claimed that it is possible to build innovative artificial molecular machines such as gears and bearings by using a universal assembler that assembles atoms [3]. Although his idea was met with much skepticism, it led to the establishment of a research field called *molecular nanotechnology*. Progress in molecular nanotechnology began after developments in organic chemistry that made possible synthesis of molecules of various shapes and functions with a wide range of application such as carbon nanotubes. In this section, we are unable to discuss the whole area of molecular nanotechnology, so we limit our discussion to DNA nanotechnology, which is one of the lines of research in this area.

DNA nanotechnology is a technique for building various nanometer scale structures and devices using nucleic acids such as DNA and RNA. It started in mid 1980s when an American chemist Nadrian Seeman thought of creating nanostructures from DNA for the purpose of protein crystallization [4, 5].

Although DNA is the most important molecule that carries genetic information, in DNA technology it is considered simply as a programmable material. The diameter of the double helical structure of DNA is 2 nm, and the length of one spiral twist is 3.4 nm (10.5 base pairs), which means that it is possible to go beyond the resolution limit of photolithography in nanostructures of precisely sequenced DNA molecules.

There are four types of nucleobases in a DNA molecule: adenine (A), cytosine (C), guanine (G), and thymine (T). As we explained in Chapter 2, two single-stranded DNA molecules bind with each other in solution to form a double helical structure (hybridization) when their base sequences are Watson-Crick complementary (A is paired with T and C is paired with G). A single-stranded DNA molecule has no definite shape and is like a thread, but a DNA molecule in the double helical structure can be considered as a straight rod up to 50 nm.

This means that, given a solution containing single-stranded DNA molecules with various base sequences, only those strands with complementary base sequences can bond together to form a helix, while others do not bond with each other. By using this property, it is possible to make DNA molecules self-assemble into desired nanostructures.

DNA molecules used for such purposes are chemically synthesized by linking bases one by one in a DNA synthesizer. Single-stranded DNA molecules that have arbitrarily specified base sequences up to about 100 bases can be synthesized easily at low cost, and recently it has become possible even to synthesize DNA sequences of tens of thousands bases.

10.2.2 Self-assembly in DNA Nanostructures

In a biological cell, a DNA molecule is a very long double helix, which does not have any specific shape. By weaving DNA molecules into a particular configuration, a nanostructure can be constructed. In order to do so, it is necessary to make DNA molecules branch off. For this purpose, DNA junctions, i.e. branching structures of DNA molecules, can be used (Fig. 10.1(a)). A *DNA tile* is a structure consisting of two short double helices joined at two junctions (Fig. 10.1(b)) [6]. The "glue" that joins the tile is single stranded regions extending from the tile, called "sticky ends". Two DNA tiles can assemble together by joining sticky ends that have complementary base sequences (Fig. 10.1.(c)).

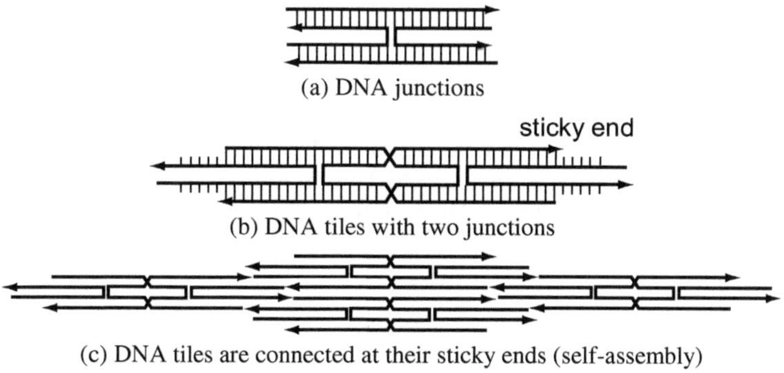

Fig. 10.1 Self-assembly of two dimensional nanostructure using DNA tiles

Since there are 4^n possible combinations for sequences of four kind of bases A, C, G and T of length n, there are quite a few sequences of sticky ends with length of five or so (1024 for a 5-base sticky end). By using DNA tiles with several different sticky ends, it is possible to make the tiles self-assemble into certain periodic or non-periodic patterns. This is called *algorithmic self-assembly* (Fig. 10.2). Generation of such tessellation patterns is closely related to the computation model of cellular automata. This process can be carried out simply by mixing DNA molecules in a test tube [7, 8].

There is also a method, called *DNA origami*, involving the folding of a long single-stranded DNA molecule. In this method, arbitrarily specified two-dimensional shapes 100 nm in length can be constructed by folding a long circular single-stranded DNA molecule consisting of about 7000 base pairs. This folding is done by attaching 250 short single-stranded DNA molecules (Fig. 10.3) [9]. This method of folding DNA molecules can be extended to three dimensions, and various three dimensional objects have been built (Fig. 10.4) [10].

Fig. 10.2 Experiment on algorithmic self-assembly of DNA tiles that generate the Sierpinski fractal pattern. Quantitative evaluation of the precision of self-assembly by comparing the atomic force microscope images of constructed DNA nanostructures with the designed cellular automata (Fujibayashi K, et al (2008) Toward Reliable Algorithmic Self-Assembly of DNA Tiles; A Fixed-Width Cellular Automaton Pattern, NanoLetters 8(7):1791-1797 ©2008 ACS)

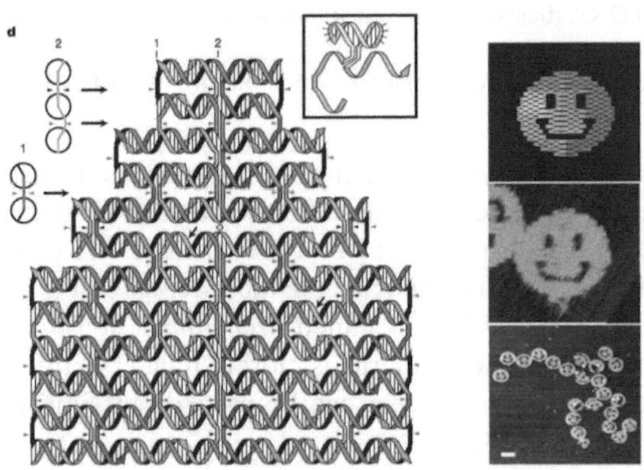

Fig. 10.3 Nanoscale structures produced by DNA origami. One long template DNA strand consisting of about 7000 bases is mixed together with 250 types of short staple DNA molecules (©2006 Nature)

Fig. 10.4 Example of three dimensional structure formed by DNA folding (octahedron) (Rothemund PWK (2006) Folding DNA to create nanoscale shapes and patterns, Nat 440:297-302 ©2004 NPG)

10.2.3 DNA Logic Gates

It is also possible to build logic gates using DNA molecules. The first proposed method employed restriction enzymes, and relatively simple finite automata have been built using the method. Afterwards, other ideas such as a logic gate utilizing only the hybridization reaction of DNA molecules, and a logic gate that uses a special sequence (called deoxyribozyme) of DNA which has restriction enzyme activity have been proposed [11].

The last of these proposed logic gates allows two hundred different types of logic gates to function in parallel in one test tube, and moreover it is possible to introduce certain relations among the inputs and outputs of these gates. This in effect amounts to constructing logic circuits. A test algorithm for tic-tac-toe has been implemented using such logic gates, which successfully demonstrated the capability of these molecules to carry out reliable computation [12].

10.2.4 DNA Sensors and DNA Actuators

For molecular scale machines, not only environmental conditions such as temperature and pH, but also existence/non-existence of various molecules can be used as possible inputs. For DNA logic gates, DNA or RNA molecules can be used as the inputs. As we explained in Section 2.4, in a biological cell, messenger RNA (mRNA) molecules are always generated in the course of gene expression. If this is used as an input, it is possible to compute certain logical operations by sensing the gene expression in progress in a cell.

There is also a particular DNA sequence called *aptamer* that changes its molecular configuration in the presence of a special target molecule so that it wraps the target molecule within [13]. It is also possible to use the different shapes of the aptamer molecule as inputs to logic gates. Utilizing this mechanism may for instance enable diagnosis of a particular disease using a complex logic system.

Furthermore, output devices, i.e. actuators, can be built entirely from DNA molecules. Fig. 10.5 shows a pair of molecular tweezers that open and close, driven by additional DNA strands called fuel [14]. Other ideas that have been proposed include a method to control hybridization using lights by inserting in the DNA sequence special bases that change their structures when lit by ultraviolet light of a particular wavelength.

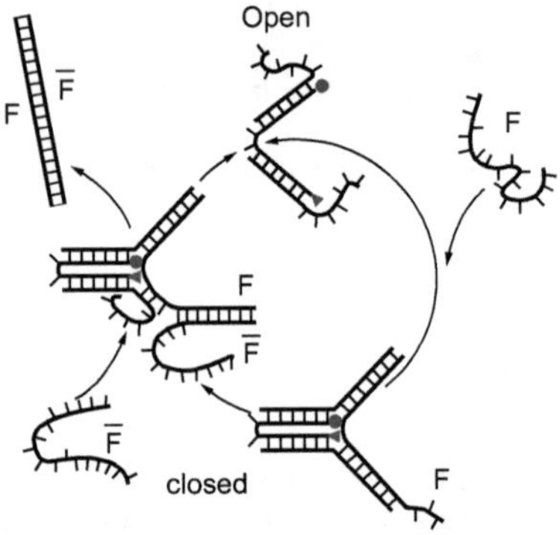

Fig. 10.5 DNA actuator. When molecule (F) is added, the tweezers close. When the complement sequence to F ($\overline{F}$) is added, the tweezers open. By alternately adding F and $\overline{F}$, these actions can be repeated any number of times (Yurke B, et al (2000)A DNA-fuelled molecular machine made of DNA, Nat 406:605-608 ©2000 NPG)

10.3 From Nanotechnology to Molecular Robotics

In the above sections we saw that DNA nanotechnology has enabled assembly of molecular scale structures, including logic gates, sensors and actuators. This implies that even robots, which have had to be built with mechatronics technology, can now be built using molecular scale devices.

Now, let us look at Fig. 1.5 in Chapter 1 once again. The horizontal axis in the graph shows the size of components, and the vertical axis shows the complexity of components on the right and the number of components on the left. Earlier we said that if the size of components is too small it is impossible for them to have any useful function, but in fact, as we saw in the real examples discussed in the previous section, all the necessary functions of robots can be carried out sufficiently by DNA molecules. This means that, even when the size of components decreases to be as small as molecules, the complexity of components does not decrease to zero, and they maintain a certain level of functional capability. On the other hand, the number of available components increases to be on the order of Avogadro's number. Consequently, the component size maximizing the functional capability of a whole system which is the product of the complexity and the number of components, is very small. (Fig. 10.6) This shows that it is no wonder that biological organisms, which are the most complex machines in nature, use components of this size range.

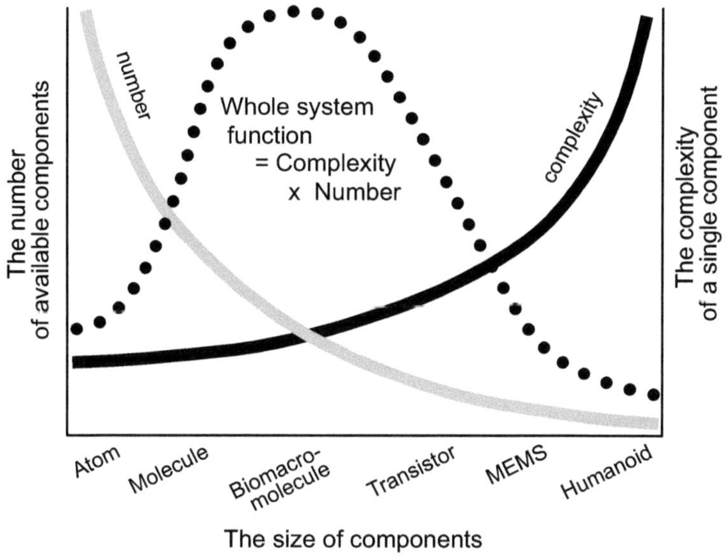

Fig. 10.6 The size of components that maximizes the functional capability of the whole system

Moreover, when molecules are used as components, there is no need for preparing a special component supply mechanism, because molecules floating in the solution are always colliding randomly with each other. This is exactly the kinetic model that von Neumann was thinking of (Section 3.1.3).

Fig. 10.7 is an illustration of an imaginary DNA robot. This is meant to be one autonomous cell whose body container, which corresponds to a cell membrane, is built by self-assembly of DNA nanostructures and is embedded with channel structures that allow transport of substances between the inside of the robot and the external environment. The channel structures are built using the three dimensional DNA Origami method and have selective permeability functions realized by DNA devices. Inside the container, DNA logic gates are stored, which are used to drive the actuators according to the type of the input molecules. This allows the DNA robot to move around in its environment.

Examples of possible applications of such a DNA robot include a drug delivery system (DDS) that delivers drugs to desired areas of a body. So-called *intracellular therapy* is also possible by using DNA robots where particular reactions are

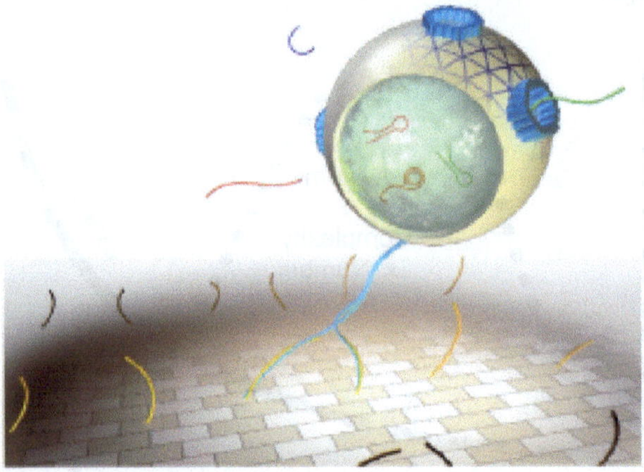

Fig. 10.7 Illustration of a DNA robot

accelerated or inhibited depending on the current expression activity of messenger RNA molecules inside a cell. Or they can be used to provide an operating platform to select and transform undifferentiated cells for regenerative medicine.

Such molecular robots with self-reproduction capability would be a step closer to realizing artificial multicellular systems that have highly advanced functions, and further on to realizing evolving artificial systems. However, in order for DNA robots to self-reproduce, DNA sequences, which are the crucial ingredients for reproduction, have to be replicated, and this will be a bottleneck. Having said that, since RNA sequences capable of RNA synthesis are already identified, it may also become possible in the future that a system consisting only of DNA molecules can replicate its own total structure while preserving the design specification (i.e., the genetic information) that it carries out[5].

It would be a great simplification of the central dogma we explained in Chapter 2 to allow DNA or RNA molecules to self-reproduce or to carry out their functions only by themselves (Fig. 10.8). Since the synthesis of protein is a very

[5] As Drexler has pointed out, endowing robots with ability to self-reproduce might make it possible for that robots to evolve by themselves independently of human intentions. The worst case scenario would be that the earth is entirely covered with self-reproducing molecular robots, a "grey goo". It is imperative that any molecular robot should be equipped with some safety mechanism to keep this from happening. Various ways to do this have actually been proposed, including making self-reproduction dependent on some special material, or installing a self-destruction device controlled by an external signal.

complex process, such a shortcut of the central dogma would improve the engineering prospects for creating a self-replicating system. All design specifications of a molecular robot built in such a way could be reduced to base sequences of A, C, G and T, and with such a robot it could be said for the first time that we have created an artificial entity that has the same information structure as biological organisms.

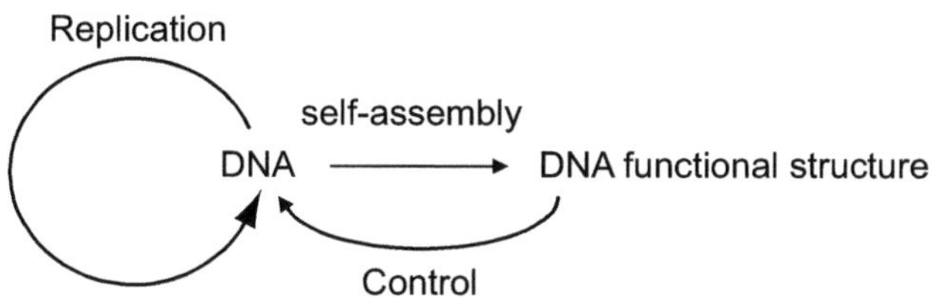

Fig. 10.8 Simplification of the central dogma

10.4 Emergence of Hierarchy: The Ultimate Problem

We have outlined a conception for constructing self-organizing mechanical systems using molecules. Even in this case, we cannot escape von Neumann's two questions (Section 3.1.1). We have to face the problem of building a reliable system using unreliable components if we are to make molecular devices. The research on how to realize computing devices such as DNA logic gates described above has begun to make progress only recently. And as for the second question of identifying the minimum logical mechanism that enables self-replication, we need to wait for further research, because DNA replication without the help of enzymes remains difficult. The authors believe, however, that these two questions can be answered.

Finally, there remains a question, the most difficult of all: how can hierarchy emerge? Let us have another look at Figure 2.1, which explains that, in biological systems, there exist natural hierarchical structures consisting of alternating layers of homogenous and heterogeneous entities. The question is why such hierarchical structures emerge in nature. We have considered various self-organizing mechanical systems from von Neumann's self-reproducing automaton to DNA robots, but none of them exhibits emergence of such hierarchy. Given a layer of components, the structure of the layer immediately above can generated by self-organization, but it is not possible to construct the layer above that.

The notion of meta-modules described in Section 7.3 introduces a kind of layer structure of modules. However, a meta-module is only a way of perceiving modules, not a result of self-organization. To solve this ultimate problem, we may have to wait until full-scale evolution experiments become possible with self-organizing robots at the molecular level. Indeed, it seems that there is no other way to approach this kind of question.

If understanding by building is the way an engineer might answer the philosophical question of why humans have evolved, research with self-organizing robots will surely lead to progress in answering that question.

References

[1] White, P.J., et al.: Stochastic Self-Reconfigurable Cellular Robotics. In: Proc. IEEE Int. Conf. Robot. Autom., vol. 3, pp. 2888–2893 (2004)
[2] Feynman, R.P.: There's plenty of room at the bottom. IEEE J. Microelectromechanical Syst. 1, 60–66 (1992)
[3] Drexler, K.E.: Engines of Creation, The Coming Era of Nanotechnology. Anchor Books, New York (1986)
[4] Chen, J., et al. (eds.): Nanotechnology: Science and Computation. Springer (2006)
[5] Komiya, K., et al.: DNA Nanoengineering, Kindaikagakusha, Tokyo (2011) (in Japanese)
[6] Winfree, E., et al.: Design and self-assembly of two-dimensional DNA crystals. Nat. 394, 539–544 (1998)
[7] Rothemund, P.W.K., et al.: Algorithmic Self-Assembly of DNA Sierpinski Triangles. PLOS Biology (2004)
[8] Fujibayashi, K., et al.: Toward Reliable Algorithmic Self-Assembly of DNA Tiles; A Fixed-Width Cellular Automaton Pattern. NanoLetters 8(7), 1791–1797 (2008)
[9] Rothemund, P.W.K.: Folding DNA to create nanoscale shapes and patterns. Nat. 440, 297–302 (2006)
[10] Shih, W.M., et al.: A 1.7-kilobase single-stranded DNA that folds into a nanoscale octahedron. Nat. 427, 618–621 (2004)
[11] Stojanovic, M., Stefanovic, D.: A deoxyribozime-based molecular automaton. Nat. Biotechnology 21(9), 1069–1074 (2003)
[12] Pei, R., et al.: Training a molecular automaton to play a game. Nature Nanotechnol. 5, 773–777
[13] Ellington, A.D., Szostak, J.W.: In vitro selection of RNA molecules that bind specific ligands. Nat. 346(6287), 818–822 (1990)
[14] Yurke, B., et al.: A DNA-fuelled molecular machine made of DNA. Nat. 406, 605–608 (2000)

Subject Index

absolute coordinate system 84
activation criteria 90
activator 29, 67
activation 90
actuator 2
adaptability 15
adenine 22
adjacent module communication 221
adjacent type list 87
algorithmic self-assembly 239
amino acid 23
anchor 156
anchor formation 156
angular error 109
animal herd 34
anonymity 61
ant 33
aptamer 241
artificial life 48
artificially synthesized molecules 237
Astro Boy 16
asynchrony 61
atom 19
ATRON 121
automaton 38
autonomy 14
Avogadro's number 15

base state 134
baseline design 4
bee 33
Belousov-Zhabotinsky chemical reaction (BZ reaction) 68
biological development 21
biological organisms 237
biological self-repair 30
biological system 19
biological tissue 116
block models of self-reproducing machine 45
bluetooth 220
brain 16, 38
bus communication 221
butterfly effect 6
Byzantine agreement problem 75
Byzantine generals problem 75

C. elegans 26
cascading 28
CEBOT 110
cell 19
cell lineage tree 26
cellular automaton, cellular automata 38, 69, 102, 117, 159, 239
cellular model 42
cellular slime mold 33
central dogma 24
Central Pattern Generator (CPG) 193
chain-type 106
Chirikjian, Gregory 115
CHOBIE 116
Claytronics 126
cluster flow 151
cockroach legs 31
coincidence organ 41
collision avoidance 136, 162
compensatory regeneration 31
complementary shape 109
completeness 96
complexity 15, 41
component 14, 78
computer simulation 91
conceptual design 4
configuration recognition 226
conformation 15
connection mechanism 109, 214
connection type 86
connectivity 78
connector 109
CONRO 125
constructive approach 37

Controller Area Network (CAN) 219
cooperative phenomena 20
coordination 14
coupled oscillator 184
crawler 179
crossover 198
crystalline 116, 148
cube 108, 117
cutting organ 42
cyclic AMP 33
cytosine 22

deadlock 74, 91, 162
deadlock avoidance 162
debugging 14
degrees of freedom 108
deoxyribozyme 241
dependability 13
description 85
description matrix 93, 94
description tape 42
design by self-organization 7, 8
design model 5
design variable 5
detailed design 4
detection of the loss 98
difference measure 88
diffusion 61, 96
diffusion field 88, 90
Digital Mirror Device (DMD) 15
dissipative structure 8
distance between connection types 87
distributed algorithm 72
distributed autonomous system 8, 14, 77
distributed metamorphosis 230
disturbance estimation 206
divergence 63
DNA 22
DNA actuator 241
DNA logic gates 241
DNA nanostructure 239, 243
DNA nanotechnology 238
DNA origami 239
DNA polymerase 25
DNA sensor 241
DNA synthesizer 238
DNA tile 239
double helix 22
Drexler, K. Eric 238, 244
drug delivery system (DDS) 243
duplication of components 13

dynamic walk 176
dynamics simulation 200

EDVAC 38
eigenvalue 64
emergence of hierarchy 245
engineering to do something 7
engineering to let things become something 7
entrainment 187
environment 2
epimorphosis 31
evaluation criteria 5
exclusion control 74
exhaustive search 140
extensibility 12

failsafe design 13
fault-tolerance, fault-tolerant 12–13
fault-tolerant design 12, 75
FDM (Fused Deposition Modeling) 116
Feynman, Richard 238
fertilization 84
field of interaction 14
finite element analysis 117
fitness 200
fixed configuration 105
flexibility 11
flow field 63
foolproof 13
formal language theory 38
forward problem 5
Fractal Machine 113
Fractum, Fracta 77, 81
Fukuda, Toshio 110
fusing organ 42

gait 176
game of life 71
gene 198
generation 198
Genetic Algorithm (GA) 197, 198
genetic information 22
germ layer lineage 27
global consensus 162
global entrainment 194
global information 14
global order 14
graceful degradation 13
gradient field 63
graph automata 50
Griffith, Saul 56

grey goo 244
Goldstein, Seth 126
guanine 22

Haken's synergetics 8
herd 19
hermaphrodite 109
hermaphrodite connector 119
heterogeneous 59
heuristics 141
hexagonal lattice 81
hexagonal lattice-type module 115
hierarchical system 19
homogeneity 60, 77
homogeneity of individuals 14
homogeneous 59
homogenous system 77
Hosokawa, Kazuo 52
human error 13
humanoid 16
hybrid type 128
hybridization 20, 22, 238
hydrogen bond 22

ID number 83
individual 8, 14, 19
information processing component 2
inhibitor 29, 67
inhibitory organ 41
in-lattice state 132
Inou, Norio 116
input-output characteristics 39
input-output correspondence 2
integral compensation 206
INTEL 4004 15
Internally Balanced Magnetic Unit (IBMU) 214
Internet 12
inter-unit communication 83
intestinal villi 30
intracellular therapy 243
inverse kinematics 174
Ishiguro, Akio 127

Kelvin's fourteen-faced polyhedra 108
kernel 93
kinematics 174
Klavins, Eric 55
Kokaji, Shigeru 112
Kondo, Shigeru 29

Langton, Christopher 49
Langton's loop 50
lattice-type 106
lattice-type modular robot 131
leader election 73, 83, 226
leak constant 90
LEGO MINDSTORMS® 105
lethal mutation 41
limit cycle 188
limited space 107
lipid 20
Lipson, Hod 123
local communication 77
local entrainment 195
local interaction 59
locality 60
locality of communication 14
localization 114
location index 94
locomotion 114
logical connection 94
logical organization 38
logical type 94
Lund, Henrik 121

machines composed of biopolymers 237
magnet 81
magnetic unit 52
manually reconfigurable modular robot 105
Markov process 56
Matsuoka's oscillator 195
McCulloch, Warren 39
mechanical system 2
mechatronic self-assembling system 55
mechatronics 235
messenger RNA (mRNA) 25
meta-design 7
meta-module 146, 245
Metamorphic Robot 115
metamorphosis 131
metamorphosis procedure 134
Micro Module 116
micromachine 15
minimal completeness propagation algorithm 96
mobility 78, 108
mode conversion 136
modular robot 105

module 11
module size 235
Molecube 123
molecular machine 19, 237
molecular nanotechnology 238
molecular robotics 242
molecular tweezers 242
Molecule 120
molecule 19
Moore neighborhood 69
Moore's Law 235
morphogen 27
morpholaxis 31
movability 89
movable type 88, 89
mRNA 25
M-TRAN 129, 131
multicellular organization 25
multiple-DOF 113
multiple-DOF linkage structure 114
muscle 42
mutation 41, 198
mutual exclusion 74
MX2 112

N-point crossover 198
neural oscillator 195
neuron 16
non-determinism of interactions 14
non-linear property 6
NP-hard 141
nucleic acid 20
number of components 15
number of modules 236

one-to-one correspondence of functions and components 4
onion method 93, 94
open system 14
optimization 198
order formation 14
ordinary transmission cell 42
organ 19
organelle 19
origin of assembly 84

Penrose's block 38
Penrose, Lionel S. 45
phase 184
phase oscillator 185
phase plane 188
photolithography 236
physical types 94
physiological regeneration 30
pivoting 134
Pitts, Walter 39
planaria 31
polarity 138
PolyBot 125
polymer 22
PolyPod 124
porous structure 165
power source 2
preconditions for self-assembly algorithm 83
Prigogine, Ilya 8
principle of minimization of free energy 20, 23
process of evolution 40
programmable material 238
protein 20, 23
protein folding 23
proximity sensors 212

quantitative analysis of self-reproducing system 52

RAID system 13
random coil 23
ratchet 47
reaction-diffusion model 29
reaction-diffusion system 8, 63
reconfiguration 78
reconstruction 30
reductionism 1
reductionist design 1, 4
redundancy 75
reflex control 195
regular structure 146, 148
reinforcement learning 208
relative position error 109
reliability 74
reliability engineering 39
replication 25
restriction enzyme 241
retrogression of the stage 98
retrogression signal 98
rhombic dodecahedra 108
ribosome 23
rigid member 42
RNA 22
rolling 134
rotational motion 109
Rus, Daniela 116, 120

Sanderson, Arthur C. 114
scalability 12
Seeman, Nadrian 238
selection 198
self-assembly 77, 79, 85
self-assembly problem 85
self-complementary shape 109
self-healing 21
self-organization 7, 9
self-organizing machine 37
self-organizing robot 235
self-reconfigurable modular robot 106
self-reconfiguration 131, 138
self-repair 80
self-replicable modular robot 106
self-replication 168
self-reproducing automata 39
self-reproducing molecular
robots 244
self-reproducing system 56
self-reproduction 21, 38, 44, 45, 123
self-reproduction of a Turing
machine 51
sensor 2
separation avoidance 136
shape memory alloy (SMA) 116, 213
Sierpinski gasket 113
sigmoid function 197
simulation 5
simulation of self-assembly 97
simulation of self-repair 98
simulator 223
Shen, Wei-Min 125
SlimeBot 127, 192, 232
SMA actuator 125
social insect 33
society 19
solution trajectory 188
space filling tessellation 108
space-filling polyhedra 108
spanning tree 74
spatio-temporal symmetry
breaking 84
special transmission cell 43
specification description language 13
spontaneity 14
stage 94
staged self-assembly and
self-repair 92
statement 87
static walk 176
sticky end 239
stimulus organ 41
stimulus producer 41
strategy for self-assembly 88
stress 117
structural component 2
structure rewriting rule 51
sugar chain 20
support leg 175
swing leg 175
symmetric transformation 228
symmetry 108
symmetry breaking 84
synchronization 114, 184

target configuration 85
target type 87
telescopic link mechanism 113
Terminator 2 79, 127
tessellation 108
TETROBOT 114
theory of self- reproducing
automata 38
there is plenty of room at the
bottom 238
thermal agitation 237
three dimensional universal
connection system 117, 131
thymine 22
time origin 84
topology 5
totipotent 26
transcription 25
transition diagram 87
translation 25, 109
traveling wave 180
tricolor flag model 27
true regeneration 31
truncated octahedra 108
Turing, Alan 29
Turing instability 65
Turing pattern 67
Turing's universal automata 39
two dimensional unit 80

UNDX (unimodal normal distribution
crossover) 198
unit 78
unit identifier 83
unit supply by circulation 97
universal assembly 138
universal automaton, universal
automata 39

van der Pol oscillator 197
Velcro 127
von Neumann, John 37, 78
von Neumann architecture 38
von Neumann neighborhood 69
von Neumann's two questions 245

wake-up 114
walker generation 155
walker pair 155
Watson-Crick complementary 22, 238
Wolpert, Lewis 27
World Wide Web (WWW) 6, 16
wound healing 31

YaMoR 197
yield 54
Yim, Mark 124

zygote 25

Springer Tracts in Advanced Robotics

Edited by B. Siciliano and O. Khatib

Further volumes of this series can be found on our homepage: springer.com

Vol. 77: Murata, S.; Kurokawa, H.
Self-Organizing Robots
252 p. 2012 [978-4-431-54054-0]

Vol. 76: Prassler, E.; Bischoff, R.; Burgard, W.; Haschke, R.; Hägele, M.; Lawitzky, G.; Nebel, B.; Plöger, P.; Reiser, U.; Zöllner, M.
Towards Service Robots for Everyday Environments
528 p. 2012 [978-3-642-25115-3]

Vol. 75: Civera, J.; Davison, A.J.; Montiel, J.M.M.
Structure from Motion Using the Extended Kalman Filter
168 p. 2012 [978-3-642-24833-7]

Vol. 73: Corke, P.;
Robotics, Vision and Control
XXX p. 2011 [978-3-642-20143-1]

Vol. 72: Mullane, J.; Vo, B.-N.; Adams, M.; Vo, B.-T.
Random Finite Sets for Robot Mapping and SLAM
146 p. 2011 [978-3-642-21389-2]

Vol. 70: Pradalier, C.; Siegwart, R.; Hirzinger, G. (Eds.)
Robotics Research
752 p. 2011 [978-3-642-19456-6]

Vol. 69: Rocon, E.; Pons, J.L.
Exoskeletons in Rehabilitation Robotics
138 p. 2010 [978-3-642-17658-6]

Vol. 68: Hsu, D.; Isler, V.; Latombe, J.-C.; Ming C. Lin (Eds.)
Algorithmic Foundations of Robotics IX
424 p. 2010 [978-3-642-17451-3]

Vol. 67: Schütz, D.; Wahl, F.M. (Eds.)
Robotic Systems for Handling and Assembly
460 p. 2010 [978-3-642-16784-3]

Vol. 66: Kaneko, M.; Nakamura, Y. (Eds.)
Robotics Research
450 p. 2010 [978-3-642-14742-5]

Vol. 65: Ribas, D.; Ridao, P.; Neira, J.
Underwater SLAM for Structured Environments Using an Imaging Sonar
142 p. 2010 [978-3-642-14039-6]

Vol. 64: Vasquez Govea, A.D.
Incremental Learning for Motion Prediction of Pedestrians and Vehicles
153 p. 2010 [978-3-642-13641-2]

Vol. 63: Vanderborght, B.;
Dynamic Stabilisation of the Biped Lucy Powered by Actuators with Controllable Stiffness
281 p. 2010 [978-3-642-13416-6]

Vol. 62: Howard, A.; Iagnemma, K.; Kelly, A. (Eds.):
Field and Service Robotics
511 p. 2010 [978-3-642-13407-4]

Vol. 61: Mozos, Ó.M.
Semantic Labeling of Places with Mobile Robots
134 p. 2010 [978-3-642-11209-6]

Vol. 60: Zhu, W.-H.
Virtual Decomposition Control – Toward Hyper Degrees of Freedom Robots
443 p. 2010 [978-3-642-10723-8]

Vol. 59: Otake, M.
Electroactive Polymer Gel Robots – Modelling and Control of Artificial Muscles
238 p. 2010 [978-3-540-23955-0]

Vol. 58: Kröger, T.
On-Line Trajectory Generation in Robotic Systems – Basic Concepts for Instantaneous Reactions to Unforeseen (Sensor) Events
230 p. 2010 [978-3-642-05174-6]

Vol. 57: Chirikjian, G.S.; Choset, H.; Morales, M., Murphey, T. (Eds.)
Algorithmic Foundations of Robotics VIII – Selected Contributions of the Eighth International Workshop on the Algorithmic Foundations of Robotics
680 p. 2010 [978-3-642-00311-0]

Vol. 56: Buehler, M.; Iagnemma, K.;
Singh S. (Eds.)
The DARPA Urban Challenge – Autonomous Vehicles in City Traffic
625 p. 2009 [978-3-642-03990-4]

Vol. 55: Stachniss, C.
Robotic Mapping and Exploration
196 p. 2009 [978-3-642-01096-5]

Vol. 54: Khatib, O.; Kumar, V.;
Pappas, G.J. (Eds.)
Experimental Robotics:
The Eleventh International Symposium
579 p. 2009 [978-3-642-00195-6]

Vol. 53: Duindam, V.; Stramigioli, S.
Modeling and Control for Efficient Bipedal Walking Robots
211 p. 2009 [978-3-540-89917-4]

Vol. 52: Nüchter, A.
3D Robotic Mapping
201 p. 2009 [978-3-540-89883-2]

Vol. 51: Song, D.
Sharing a Vision
186 p. 2009 [978-3-540-88064-6]

Vol. 50: Alterovitz, R.; Goldberg, K.
Motion Planning in Medicine: Optimization and Simulation Algorithms for Image-Guided Procedures
153 p. 2008 [978-3-540-69257-7]

Vol. 49: Ott, C.
Cartesian Impedance Control of Redundant and Flexible-Joint Robots
190 p. 2008 [978-3-540-69253-9]

Vol. 48: Wolter, D.
Spatial Representation and Reasoning for Robot Mapping
185 p. 2008 [978-3-540-69011-5]

Vol. 47: Akella, S.; Amato, N.;
Huang, W.; Mishra, B.; (Eds.)
Algorithmic Foundation of Robotics VII
524 p. 2008 [978-3-540-68404-6]

Vol. 46: Bessière, P.; Laugier, C.;
Siegwart R. (Eds.)
Probabilistic Reasoning and Decision Making in Sensory-Motor Systems
375 p. 2008 [978-3-540-79006-8]

Vol. 45: Bicchi, A.; Buss, M.;
Ernst, M.O.; Peer A. (Eds.)
The Sense of Touch and Its Rendering
281 p. 2008 [978-3-540-79034-1]

Vol. 44: Bruyninckx, H.; Přeučil, L.;
Kulich, M. (Eds.)
European Robotics Symposium 2008
356 p. 2008 [978-3-540-78315-2]

Vol. 43: Lamon, P.
3D-Position Tracking and Control for All-Terrain Robots
105 p. 2008 [978-3-540-78286-5]

Vol. 42: Laugier, C.; Siegwart, R. (Eds.)
Field and Service Robotics
597 p. 2008 [978-3-540-75403-9]

Vol. 41: Milford, M.J.
Robot Navigation from Nature
194 p. 2008 [978-3-540-77519-5]

Vol. 40: Birglen, L.; Laliberté, T.; Gosselin, C.
Underactuated Robotic Hands
241 p. 2008 [978-3-540-77458-7]

Vol. 39: Khatib, O.; Kumar, V.; Rus, D. (Eds.)
Experimental Robotics
563 p. 2008 [978-3-540-77456-3]

Vol. 38: Jefferies, M.E.; Yeap, W.-K. (Eds.)
Robotics and Cognitive Approaches to Spatial Mapping
328 p. 2008 [978-3-540-75386-5]

Vol. 37: Ollero, A.; Maza, I. (Eds.)
Multiple Heterogeneous Unmanned Aerial Vehicles
233 p. 2007 [978-3-540-73957-9]

Vol. 36: Buehler, M.; Iagnemma, K.;
Singh, S. (Eds.)
The 2005 DARPA Grand Challenge – The Great Robot Race
520 p. 2007 [978-3-540-73428-4]

Vol. 35: Laugier, C.; Chatila, R. (Eds.)
Autonomous Navigation in Dynamic Environments
169 p. 2007 [978-3-540-73421-5]

Vol. 34: Wisse, M.; van der Linde, R.Q.
Delft Pneumatic Bipeds
136 p. 2007 [978-3-540-72807-8]

MIX
Papier aus verantwortungsvollen Quellen
Paper from responsible sources
FSC® C105338

If you have any concerns about our products,
you can contact us on
ProductSafety@springernature.com

In case Publisher is established outside the EU,
the EU authorized representative is:
Springer Nature Customer Service Center GmbH
Europaplatz 3, 69115 Heidelberg, Germany

Printed by Libri Plureos GmbH
in Hamburg, Germany